An Introduction to Food Rheology

An Introduction to
Food Rheology

H. G. MULLER
M.Sc., Ph.D. (Lond.), F.I.F.S.T.

Proctor Department of Food and Leather Science
The University of Leeds

CRANE, RUSSAK & COMPANY, INC
NEW YORK

First published in the United States 1973 by:
Crane, Russak & Company, Inc
52 Vanderbilt Avenue
New York, N.Y. 10017

Library of Congress Catalog No. 72-81532
ISBN 0-8448-0013-9

Printed in Great Britain

Geh hin, mein Buch, in alle Welt;
Steh aus, was dir kommt zu.
Man beisse dich, man reisse dich;
Nur dass man mir nichts tu.

FRIEDRICH FREIHERR VON LOGAU,
1604–55

[Be off, my book and see the world.
Travel far over land and sea.
If they tear you to pieces, I don't care,
Provided they don't do it to me.]

Preface

In the food industry a considerable amount of rheological testing is carried out and much of it can be improved. Strangely enough, there is also an abundance of books on rheology but the information is often not utilized in practice. The reason, I think, lies in the fact that these texts are too mathematical for the average food scientist.

One of my students once said: 'To understand rheology one has to learn a lot of mathematics—or take a different point of view'.

I have taken that different point of view.

This book is meant to teach food rheology to the student and food technologist in industry without the use of calculus or mathematics beyond school 'O' level. This is how it has been done:

The theoretical rheologist supplies the ideal models which the applied rheologist uses to approximate to his real materials. The behaviour of these models can be expressed either by diagrams or by mathematical equations. The latter provide great precision, but what point is there for such precision if no real food resembles these models at all closely? So diagrams are quite adequate.

This leaves us with the mathematics underlying measuring techniques. If we apply the rule that equations, like books or concubines, are for use rather than ostentation, more mathematics can be dispensed with. There remain the strictly useful and surprisingly simple equations indispensable to rheological measurement. These are explained in the text. Examples and a practical section in the appendix are given to make their significance quite clear.

So the book has been written for the literate rather than the numerate, for the experimentalist rather than the theoretician. It should give the reader an advantage if he wishes to attempt the rheological analysis of any food product.

I am grateful to the following who have kindly given permission for the reproduction of diagrams or tables: Professor P. Cox; Professor A. G. Ward; Professor P. F. Pelshenke; Dr P. Søltoft; Mrs P. Goodman; Dr P. Wade; Mr R. H. McDowell; Dr R. BeMiller; Dr H. Brüning; the American Society of Agricultural Engineers; the U.S. Bureau of Standards; the Editor of the *Journal of Agricultural Research*; John Wiley & Sons, Inc; Academic Press, Inc; Pergamon Press, Ltd; the American Oil Chemists Society; North Holland Publishing Co; the Institute of Food Technologists; Masson & Cie, Paris; the American Association of Cereal Chemists; the American Chemical Society; and the Institute of Food Science and Technology of the United Kingdom.

I wish to thank sincerely my colleagues and students for their advice and criticism. Particularly I wish to thank the head of my department, Professor A. G. Ward, and my colleague Jack Lamb. They gave generously of time and knowledge and without their support this book would not have appeared in its present form. Finally I wish to thank my wife, Helen, as much for her technical assistance as for her moral support. I dedicate this book to her.

Leeds   H. G. MULLER

Contents

1 Introduction

1.1 THE PSYCHOLOGICAL AND RHEOLOGICAL APPROACH TO MECHANICAL BEHAVIOUR

Just as foods have colour, smell, and taste, so they exhibit mechanical behaviour: they react in certain ways when we try to deform them. They may be hard or soft, tender or tough, chewy or brittle, smooth or stringy. They may flow easily or with difficulty.

We can assess the mechanical behaviour of food in two ways. We can touch the food, squeeze it, bite it or chew it and say what we think of it. This is the sensory (physiological/psychological) approach. Peoples' reactions vary and so statistics will be required. The food can be evaluated by sensory panels which may place the food within certain rating or preference scales. This approach to mechanical behaviour has been called 'Haptaesthesis' (Greek: touch sensation)[1].

Haptaesthesis is a branch of psychology (or alternatively of sensory physiology) and deals with the perception of the mechanical behaviour of materials through the senses.

The second is the physical approach. This is independent of the individual, uses instruments and has been called 'objective'. Results are returned in terms of metres (m), kilograms (kg), and seconds (s). It is with this aspect, that this book is primarily concerned. The physical approach to mechanical behaviour is called rheology. Rheology is a branch of physics and is defined as the science of deformation of matter.

The interaction between the haptaesthetic and rheological approaches has been referred to as psychorheology[2]. It is the study of the relationship between subjective assessment and rheological measurement.

1.2 RHEOLOGY

Rheology, the science of deformation of matter[3] deals mainly with the deformation of apparently continuous and coherent bodies, but friction between solids, the flow of powders or even particle reduction such as milling, emulsification or atomization are often included.

(a) Reasons for rheological study

There are four main reasons for the study of the rheological behaviour of materials. First, such study allows an insight into structure. For instance, there is a relation between the molecular size and shape of materials in solution and their viscosity. There is a relation between the degree of cross linkage of polymers and their elasticity. Secondly, rheological testing is often used in raw material and process control in industry. The rheological control of wheat flour dough during bread manufacture is an example.

Thirdly, rheology has applications to machine design. Pumps, pipe-lines, and hoppers must be suitable for the material for which they are to be used. Engineers always allow for a 'factor of safety'. 'Factor of ignorance' is perhaps a better term: it costs money. The greater the knowledge of the rheology of the material, the more efficient are the pumps or the hopper.

Fourthly, rheology has relevance to acceptability by the consumer. Rheological evaluation of the spreadability of margarine, the viscosity of a milk shake, the hardness of a boiled sweet, and the toughness of meat are all examples of this aspect.

(b) Difficulties of rheological evaluation

There are essentially two difficulties in rheological classification. The first is the enormous range of materials: there are liquids, solids, gases, and materials with intermediate rheological properties. We find foods such as milk, mayonnaise, cream cheese, bread crumb, and peppermints. How can such a variety be classified rheologically?

The second difficulty is that any material behaves differently under different conditions. If a stone falls on your head, the stone acts as a solid, but if we observe the folding and settling of geological strata, the stone acts as a liquid. If we tap a window pane, it is solid, but if we measure the thickness of a very old window glass, we shall find that it is thicker at the bottom than at the top; during the centuries it has flowed like a liquid. Whether a body behaves as a solid or a liquid depends on how long a force is applied to it or how large the force is.

A 'Deborah number' has been defined. It is the time of observation divided by the time of deformation (or strictly 'relaxation') (*see* Chapter 10). The greater the Deborah number the more solid the material appears; the smaller the Deborah number, the more liquid it appears.

Deborah was a prophetess of the Old Testament and she sang:

> Lord, when thou wentest out of Seir,
> when thou marchedst out of the field of Edom,
> the earth trembled, and the heavens dropped,
> the clouds also dropped water.

> The mountains flowed before the Lord,
> *even* that Sinai from before the Lord God of Israel[4].

So, on God's infinite time scale even the mountains will eventually flow before you: hence the Deborah number. But even under normal conditions of everyday life, all materials can exhibit all rheological properties, liquid and solid, although some predominate[3].

So there are two difficulties in the rheological classification of materials: first an almost infinite variety of materials and secondly each one of these materials has different properties under different conditions.

How has the rheologist solved these two problems?

1.3 RHEOLOGICAL REFERENCE MATERIALS

To deal with the large range of materials, the rheologist specifies ideal reference materials. If all solids, (rock, steel, glass) exhibit some liquid properties, there is one that does not, and that is the ideal solid. It does not show any liquid properties because we define it so. Similarly the ideal liquid, by definition, does not show any solid properties. It is the liquid *par excellence*. The ideal solid is called the Hooke solid after Robert Hooke (1635–1705), an English inventor and architect. (Apart from concerning himself with elasticity, he was the first to apply the spring-balance wheel to watches and he designed the Montague House and the College of Physicians in London[5].)

The ideal liquid is named after Sir Isaac Newton (1642–1726) an eminent English natural philosopher and mathematician.

The Hooke solid (*see* Chapter 2) and the Newton liquid (*see* Chapter 4) form the very limits of rheological behaviour. Whatever strange materials we may find in real life, none is more solid than the Hooke solid and none more liquid than the Newtonian liquid. Both are structureless (there are no atoms), both are isotropic (they have the same properties in all directions), and they follow their respective laws exactly. They are ideal materials and of course like ideal husbands or wives, they don't exist in reality.

Later we shall see how by combining these two ideal rheological models we obtain intermediate ones: for example, the Bingham model (*see* Chapter 8) represents an ideal plastic material and the Maxwell and Kelvin-Voigt models (*see* Chapter 10) represent the ideal visco-elastic liquid and solid respectively. These models are used by the rheologist much as the geographer uses his lines of latitude and longitude, which are equally imaginary, to find his way through a real and complicated world.

The second problem is that the same material will behave differently under different conditions. How can this be solved? To overcome it, the rheologist specifies the conditions of his test in m, kg, and s, as well as the temperature of his material. Experience shows that there are certain reasonable and practical test conditions. Under these the elasticity of water for instance, becomes negligible and so does the viscosity of rock.

We must now consider these testing conditions somewhat more closely.

1.4 FORCE AND DEFORMATION

(a) *Force, traction, and stress*

A force, although defined in terms of its power to produce acceleration, is also an agency capable of deforming a material. For convenience it is given the symbol 'F'. Force is not a useful rheological criterion. If I sit on a chair, all is well. If I sit on a pin, all is not well. The force with which I press down (the load) is still the same, but the area of application has greatly decreased. Therefore it is more meaningful to use force (F) divided by the unit area (A). Force divided by unit area is called 'traction'. The unit of force is the newton and the unit area the square metre. Hence traction is given in newtons per square metre (N/m^2). One newton is that force which will give an acceleration of 1 m/s/s to a mass of 1 kg. In the older literature 'dyne' is often used as the unit of force, which gives 1 g an acceleration of 1 cm/s/s. One newton equals 100,000 (or 10^5) dynes.

The newton is really quite a small unit. It is roughly the force experienced when holding a 100 g weight in the outstretched hand.

What is stress? Imagine a uniform cylinder of a solid material being pulled by a force (*see* Figure 1.1) which applies a traction to the

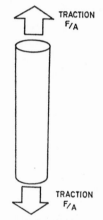

TRACTION
F/A

TRACTION
F/A

Figure 1.1 Extended cylinder in state of stress

end surface. It will be in tension. Should you cut the body across, it would fly apart. The traction you would have to apply to bring the cut surfaces together again is equal to the original traction applied. The bottom part of the cylinder fulfilled that function before the cut was made. The cylinder in tension (or compression) is said to be in a 'state of stress' which in this case is uniform throughout its interior. As the cylinder is extended it contracts laterally: it becomes narrower and its shape is changed. The state of stress is a three-dimensional phenomenon most simply described as above, but stress components can be demonstrated in various directions. By a careful choice of experimental conditions (choice of force, small deformations only, and geometry of sample) this complex stress situation may be simplified to enable only one stress component to be considered. In the above experiment the traction applied is equivalent to the longitudinal tensile stress component. For a consideration of detailed three dimensional stress analysis the reader should consult the more advanced literature[6,7].

(b) *Deformation and strain*

If a traction acts on a body, if a body is in a state of stress, it undergoes deformation. Extension or compression are examples. A more useful measure than deformation is 'relative deformation'. This is the change in dimension experienced by each unit of dimension. Imagine for instance a wire being stretched. If the original length is L and the increase in length l (its final length thus being $L + l$) then l/L is the relative deformation or 'strain'. The same strain would be obtained for every portion of the wire.

Thus a given stress causes a strain, but the magnitude of the strain depends on the type of material. For the same stress some metal wires for instance show larger strains than others. Rubber shows large strains for quite small stresses.

For many materials, and with modest strains, experiment shows that a stress numerically equals the resultant strain multiplied by a constant, i.e. stress and strain are proportional to one another. That constant is called an elastic modulus.

$$\text{Stress} = \text{Strain} \times \text{Modulus}$$

This equation is typical of an elastic solid. It is called an equation of state.

(A rheological equation of state is the mathematical relationship between stress and strain or rate of strain.)

Time does not enter into this equation. For instance, if a copper wire is loaded with a weight, we can measure its deformation without requiring a clock. In fact we can load the wire, busy ourselves elsewhere and come back half an hour later and take the reading. It is true that when the load is added the strain is not instantaneous, but the time needed is so short that special techniques would be required to measure it.

(c) *Flow and rate of strain*

Let us now consider deformation in liquids which is called 'flow'. We fill two identical funnels, one with syrup and the other with water. Then we go away for a while and when we return both funnels are empty: to determine the rheological difference between syrup and water a clock is required. In elasticity experiments we deal only with the *magnitude* of deformation, in viscosity experiments we are concerned with the magnitude of deformation divided by the time taken for it, in other words the *rate* of deformation. Viscous flow is a rate process.

As with solids, there is also an equation of state for liquids. With a solid, the rheological equation of state connects stress and strain, with a liquid it connects stress with rate of strain. The elastic constant was called a 'modulus', the viscous constant is called the coefficient of viscosity. Hence for a liquid the equation derived by experiment states that for many real materials:

$$\text{Stress} = \text{Rate of strain} \times \text{Viscosity coefficient}$$

Although for very viscous liquids, like pitch, a cylinder could be made to extend by flowing, we shall see later that the rate of strain in a liquid can most easily be expressed as rate of shear, the corresponding stress being a shearing stress (*see* later).

(d) *Empirical testing*

Fundamental rheological test results are returned in terms of kilograms, metres, and seconds. Whatever the method used, the same result is obtained within the experimental error. Reducing all results to these three basic units is of course a great advantage. Not only are they easily understandable but the basic units themselves are easily accessible as standards.

Unfortunately food is exceedingly complicated rheologically and fundamental testing is often laborious and time consuming and does not give simple answers. An empirical method may provide very useful information even if the results cannot be compared for different methods of test. For instance a load-extension curve is so much easier to obtain than a stress-strain curve, where cross sectional areas and the relative deformations of the specimen must be calculated. A survey in 1968 showed that almost all routine testing in the British food industry was empirical[8].

The advantage of this approach is that empirical testing is so much quicker and simpler than fundamental testing. The disadvantage is that the results are specific to a particular instrument.

Empirical testing should be strictly correlated with performance. If no such correlation can be made, the test may be of doubtful value. One must also distinguish between empirical tests and application of a faulty fundamental method. The latter pretends to be fundamental when it is not. Many examples can be found in the literature and several instances are considered in the following chapters.

2 Solids: Characteristics and Measurement

2.1 THE HOOKE SOLID

It seems that Hooke discovered the law that bears his name in 1660 and published it in the form of an anagram in 1676. It read:

c e i i i n o s s s t t u v

In 1678 he explained it: *Ut tensio sic vis*, or in translation, 'as the tension, so the force'.

The Hooke solid is the theoretical concept of a solid, where the deformation is proportional to the force (or traction) producing it. It has no viscous properties, it is structureless and isotropic. There are no time effects (no stop watch is required for testing) and it follows the equation already given in Chapter 1 which is another way of expressing Hooke's law:

$$\text{Stress} = \text{Strain} \times \text{Modulus}$$

The deformation–time curve shows that on application of load there is an immediate deformation. As soon as the load is removed the deformation disappears completely and instantaneously (*see* Figure 2.1). This diagram illustrates the definition of ideal elasticity: the immediate and complete recovery of the strain of a material on removal of stress. Many real materials are for almost all practical purposes ideally elastic if the strain does not exceed 1 per cent.

Often rheologists find it convenient to express the Hooke model diagrammatically as a spring extended by a force, or as the shorthand symbol *H* (*see* Figure 2.2).

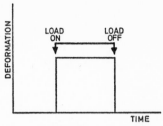

Figure 2.1 Deformation–Time curve for a Hookean solid

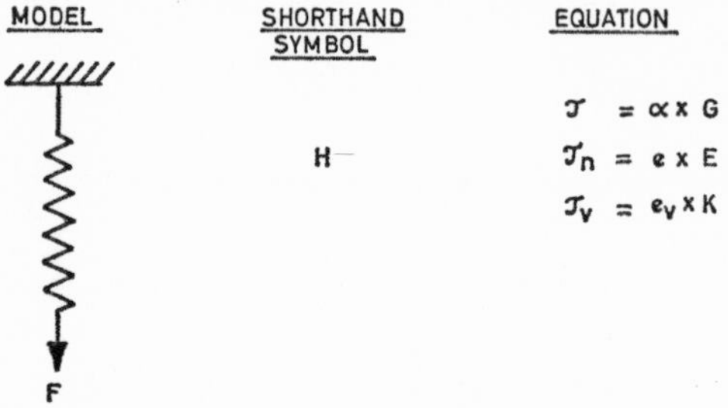

MODEL	SHORTHAND SYMBOL	EQUATION

$$\tau = \alpha \times G$$
$$\tau_n = e \times E$$
$$\tau_v = e_v \times K$$

Figure 2.2 The Hooke solid

Young's modulus

It has been pointed out in Chapter 1 that by careful choice of experimental conditions the complex three dimensional stress condition inside a body may be simplified. One such condition, the small extension or compression of a cylinder has already been described. Here a traction causes a longitudinal strain. The longitudinal strain is the deformation caused by a tensile or compressive stress. If a body is extended from length L to length $L + l$, the longitudinal strain is l/L. For this type of experiment the modulus is called Young's modulus, E, (*see* Chapter 1). Hence:

$$\tau_n = e \times E$$

where τ_n is the tensile or compressive stress and e the longitudinal strain.

The smaller the Young's modulus, the greater is the elastic deformation for a given stress, i.e. the more easily does the material deform elastically. Some Young's moduli are given in Table 2.1. Many of these materials can vary in structure and composition. This, as well as temperature differences will give rise to slightly differing values.

TABLE 2.1

Some Young's moduli (N/m^2)

Soft foam rubber	1×10^2	Concrete	$1{\cdot}7 \times 10^{10}$
Rubber	8×10^5	Glass	7×10^{10}
Wool	$0{\cdot}3 \times 10^{10}$	Iron	18×10^{10}
Dry spaghetti	$0{\cdot}3 \times 10^{10}$	Steel	25×10^{10}
Lead	$1{\cdot}0 \times 10^{10}$		

Note : To obtain results in dyne/cm^2 multiply by 10.

Shear modulus

Another experiment where the three dimensional stress condition is simplified is shear. This is the deformation caused by a shearing stress and with small deformations, the strain can be conveniently expressed as an angle. Imagine the cube in Figure 2.3(*a*) being deformed by the shearing stress τ. The amount of shear is given by the angle α, or alternatively L/H, (i.e. tan α) (*see* Figure 2.3(*b*)).

(a)   (b)

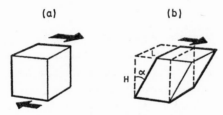

Figure 2.3 Shear of a cube

For shear, the modulus is called the shear or rigidity modulus G. Hence:

$$\tau = \alpha \times G$$

where τ is the shear stress and α the shear angle. Some shear moduli are given in Table 2.2.

It will be noted that the shear modulus is about two to three times smaller than the Young's modulus.

If isotropic, the shear and Young's moduli are all that is needed to specify an elastic material. However, for convenience, two further

TABLE 2.2

Some shear moduli (N/m²)

Liquid	0
Gelatine (80% water)	2×10^5
Rubber	$2 \cdot 9 \times 10^5$
Silk	$0 \cdot 1 \times 10^{10}$
Lead	$0 \cdot 2 \times 10^{10}$
Concrete	$0 \cdot 7 \times 10^{10}$
Glass	2×10^{10}
Copper	4×10^{10}
Iron	7×10^{10}
Steel	8×10^{10}

Note : To obtain results in dyne/cm² multiply by 10.

constants are often measured. These are Poisson's ratio and the Bulk modulus.

Poisson's ratio

When a body is stretched or compressed its width will almost always change. On compression of a cylinder its diameter will increase and on stretching, its diameter will decrease. The ratio between the lateral contraction (as a fraction of the diameter) and the longitudinal elongation or strain within the elastic range is called Poisson's ratio (μ) (Simeon Denis Poisson, 1781–1840, French mathematician). It is the same if we use metres or feet, i.e. it is dimensionless. For any material in which no volume change takes place when it is stretched or compressed (rubber approximates to this), Poisson's ratio is 0·5. For cork or breadcrumb which can be compressed without increase in diameter, Poisson's ratio is zero. Potato tissue contains less air than apple tissue, so its Poisson ratio is higher (*see* Table 2.3).

Bulk modulus

No doubt you have seen skin divers on film or television. As their exhaled air bubbles rise to the surface the bubbles increase in size. Similarly, if a body sinks in water, the pressure, caused by the weight

TABLE 2.3

Some Poisson's ratios

Cheddar cheese	0·50	Maize endosperm	0·32
Flour dough	0·50	Copper	0·30
Gelatine (80% water)	0·50	Steel	0·30
Potato tissue	0·49	Glass	0·24
Rubber	0·49	Sandstone	0·10
Shelled maize	0·40	Bread crumb	0
Apple tissue	0·37	Cork	0

of water above, compresses it. The deformation caused by the hydrostatic pressure from all sides is called volumetric strain $v/V = e_v$, where V is the original volume and v the change in volume. The modulus is called the bulk modulus K.

$$\tau_v = e_v \times K$$

In this equation τ_v is the hydrostatic pressure and e_v the volumetric strain. Table 2.4 gives some bulk moduli.

TABLE 2.4

Some bulk moduli (N/m^2)

Gas (at 1 atm.)	1×10^5	Silver	10×10^{10}
Dough	$1·4 \times 10^6$	Steel	16×10^{10}
Rubber	$1·9 \times 10^7$	Nickel	17×10^{10}
Liquid approx.	1×10^9	Diamond	$5·5 \times 10^{11}$
Granite	3×10^{10}		

Note : To obtain results in dyne/cm^2 multiply by 10.

The bulk modulus is a measure of compressibility and since all matter is to some extent compressible, all matter exhibits a bulk modulus. It is the only elastic constant for gases and liquids. For an ideal gas, the bulk modulus is not a constant but turns out to be equal to the pressure. Diamond is the least compressible substance known in nature. For that reason its bulk modulus is the highest.

Only if the material is isotropic, but only then, are the four elastic constants we have considered interconvertible. If two are known, then any of the others can be calculated. If the properties of the solid differ in different directions (if the material is anisotropic), as with crystals, many plastics, paper, and textiles, more constants are required to describe properties in the various directions. It is of course important not to confuse units of measurement, which should always be given in m and kg.

$$G = \frac{3EK}{9K - E}$$

$$K = \frac{E}{3(1 - 2\mu)} = \frac{EG}{9G - 3E} = G\,\frac{2(1 + \mu)}{3(1 - 2\mu)}$$

$$E = \frac{9GK}{3K + G} = 2G(1 + \mu) = 3K(1 - 2\mu)$$

$$\mu = \frac{E - 2G}{2G} = \frac{1 - E/3K}{2}$$

In these equations, G is the shear modulus, E the Young's modulus, K the bulk modulus, and μ Poisson's ratio.

2.2 STATIC MEASUREMENT

The Young's modulus E has been determined on rods of material such as spaghetti, apple, and potato tissues by stretching or bending the specimen.

In the stretching test, the material is rigidly clamped at the top and a weight is applied to the bottom. The extension between two marks made on the specimen is measured (*see* Figure 2.4(*a*)).

$$E = \frac{\text{stress}}{\text{strain}} = \frac{Mg/\pi r^2}{l/L} = \frac{MgL}{l\pi r^2} \; \text{N/m}^2$$

In this equation E is Young's modulus, M the load applied (kg), L the length of the specimen (m), l the extension (m), r the radius (m), g the gravitational constant (9·81), and π is 3·1416. (The mercury bath extensometer can also be used for this test—*see* Chapter 11.)

The specimen may also be clamped horizontally at one end, so that the other is free (*see* Figure 2.4(*b*)). The strain is more complex

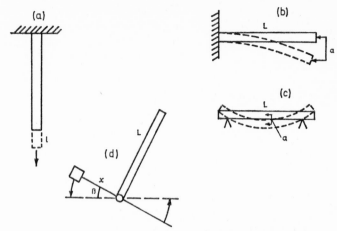

Figure 2.4 Experimental arrangements for determining
Young's modulus and shear (ridigity) modulus

and is not uniform. On loading the free end, there is a deflection,
a (m).

$$E = \frac{4MgL^3}{3\pi r^4 a} \text{ N/m}^2$$

Finally the specimen may be supported horizontally at both ends
and loaded at the centre (*see* Figure 2.4(*c*)).

$$E = \frac{MgL^3}{12\pi r^4 a} \text{ N/m}^2$$

The shear (rigidity) modulus G, is determined by fixing a rod-shaped
specimen horizontally at one end and supporting it at the other. The
free end is then twisted (*see* Figure 2.4(*d*)). Again the shear is not
uniform from outside to centre.

$$G = \frac{2Mg \times L}{\pi r^4 \beta} \text{ N/m}^2$$

The rod may also be held vertically and twisted at the bottom end.

The bulk modulus K, may be determined by placing a suitable
specimen in water in a sealed metal chamber. The force is applied by

means of compressed air and the volume change measured, using a graduated transparent tube[9] (*see* Figure 2.5).

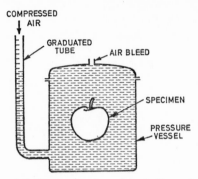

Figure 2.5 Determination of the bulk modulus[9]

This apparatus has been successfully used for fruits and tubers. For cereals, where very much larger hydrostatic pressures of up to 280 kg/cm² are required, a mechanical system is used where the liquid is compressed by a plunger.

2.3 DYNAMIC MEASUREMENT

In Figure 2.4(*b*) it is shown how Young's modulus E can be determined by loading the end of a horizontally clamped rod. One can imagine the clamped rod, say a piece of spaghetti, to be subjected to a gentle tap. The rod will vibrate and then come to rest. This vibration is referred to as 'free vibration'. Using an electric vibrator, the specimen can be subjected to forces repeatedly and at very short time intervals. This is referred to as 'forced vibration'. In this experiment the amplitude of movement can be measured. The deformation is of the same character as in the static experiment but the loading and unloading cycles are extremely fast. The time of loading (frequency) is equivalent to the time of deformation and often of the order of 50 cycles/s, i.e. the loading–unloading cycle takes one-hundredth of a second and is followed by loading and unloading in the opposite direction. Figure 2.6 shows an apple fitted with an oscillator and a detector. In this way a series of measurements can be made at different frequencies. The frequency at which maximum deformation (amplitude) is obtained, is referred to as the 'resonance frequency' which is very close to the frequency of free vibration (it will be the same if there is no damping).

So in essence, dynamic experiments are similar to static ones, except that the inertia of the specimen must be considered. Inertia is

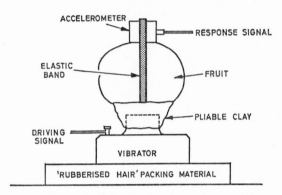

Figure 2.6 Determination of Young's modulus
using forced vibrations[10]

the tendency of a body to preserve its state of rest, or as in this case, its state of motion. So:

$$E = 4\rho f_e^2 L^2$$

where E is Young's modulus, ρ the density of the material, f_e the longitudinal resonance frequency, and L the length of the specimen. If the wavelength is very short compared with the diameter of the resonating body, a correction factor may be required.

The relationship between the shear (rigidity) modulus, G, and the frequency of a cylinder resonating in torsion is:

$$G = 4\rho f^2 L^2$$

The frequency is measured in hertz (Hz) which is equivalent to one complete cycle per second. If E and G are known, μ and K can be determined from the equations given earlier in this chapter. In dynamic experiments the elastic constants are referred to as 'dynamic elastic constants'. The results obtained are only valid if the material is homogeneous, isotropic, and shows ideal (linear) elasticity. Hence all results on food are only approximate.

In visco-elastic materials (*see* Chapter 10) damping may occur which has been expressed by the 'loss coefficient'. It gives a qualitative indication of internal viscosity (*see* Chapter 6) but the theory has not been worked out except for simple materials. Table 25 gives some dynamic Young's moduli for various fruits.

TABLE 2.5
Dynamic Young's moduli for various fruits (N/m²)

Banana	$1 - 3 \times 10^6$
Peach	approx. 10×10^6
Potato	$6 - 13 \times 10^6$
Apple	$7 - 14 \times 10^6$
Pear	$12 - 29 \times 10^6$

Note: To obtain results in dyne/cm² multiply by 10.

2.4 STRENGTH AND HARDNESS

Strength is that property of a material by which it resists either rupture or extensive deformation. There are two separate mechanisms through which a material breaks when stretched to excess and often both these mechanisms work together.

Theoretically, in an elastic solid the inter-atomic bonds support an equal share of the load. When that is exceeded all bonds snap together. In reality, the load on the sample is not distributed evenly and the bonds break in succession. Also minute flaws in the material will affect its strength. Therefore, calculated values of strength based on forces between atoms are always very much higher than those determined by experiment and the results obtained vary widely for the same sample. So a 'flaw hypothesis' has been developed which deals with the statistical distribution of the flaws in the material[11]. The number and size of flaws are determined by the structure or the treatment the material has had. The more uniform, the larger and the better aligned the structural units are, the stronger will be the material.

So with this mechanism component parts of a material will adhere to one another until the separating force becomes too great for them.

The other mechanism of rupture is based on plastic flow (*see* Chapter 8). Here the tension will force the component parts of the material to let go of their neighbours, but in doing so they cling with equal tenacity to their new ones. Thus the material exhibits plastic flow. The specimen becomes thinner if it is extended in this way and the cross sectional area on which the force acts becomes smaller, so that the stress increases. As the stress increases, the specimen deforms still faster.

Figure 2.7 shows a load extension curve and a stress–strain curve

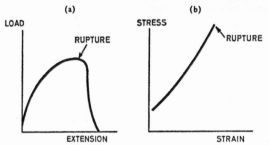

Figure 2.7 Load extension curve (*a*) and stresss–train curve (*b*) for wheat flour dough[12]

for wheat flour dough. In the former no correction for the decrease in cross sectional area is made[12].

Table 2.6 shows the breaking strengths of some common materials.

TABLE 2.6

Some breaking strengths (kg/mm²)

Liquid	0
Concrete	0·03
Spaghetti (uncooked)	1·8
Lead	1·8–2·3
Tin	2·8–3·5
Rubber	15–20
Iron	10–30
Silk	35
Copper	40
Steel	40–160

Hardness, like the storminess of the sea, is more easily appreciated than measured[13]. It is generally defined as resistance to local deformation. Hardness may be measured by pressing a square-based pyramid into the material. The indentation pressure is the load divided by the area of indentation. A spherical indenter (e.g. Brinell hardness tester) is not very suitable because indentations of various sizes formed by the same spherical indenter are not geometrically similar. A large indentation produces greater plastic strains than a smaller indentation and an appreciable rise in the observed hardness occurs. Where fracture on stretching is caused by excessive plastic deformation, hardness is usually both time dependent and related to the stress–strain curve. Often the indentation pressure is roughly three times the yield stress. This is the stress at which the sample deforms plastically during extension.

2.5 HIGH ELASTICITY

Elasticity is a characteristic of all solids such as wood, iron or concrete. It is due to the deformation of inter-atomic bonds. Young's modulus E, is of the order of 10^{10}–10^{11} N/m². If the material is stretched beyond two or three per cent, it will break.

A different type of elasticity, 'high elasticity' is found in rubber. Rubber may be stretched several hundred per cent before it breaks. Young's modulus is minute compared with that of normal solids, i.e. about 10^5 N/m². Since high elasticity was first described for rubber it is sometimes referred to as 'rubber like elasticity'[14]. It has been suggested that a material must have the following three characteristics in order to exhibit high elasticity:

1. It consists of long chain molecules with freely rotating links.
2. The secondary forces between molecules must be weak.
3. At a few points in the long chain molecule there must be firm cross links with other molecules.

The result is a large, continuous, three dimensional network, shown diagrammatically in Figure 2.8.

This concept of the structure of a rubber like material is useful, because both the number of cross links per unit volume and the molecular weight between the cross links of the material can be calculated.

It has been shown that gelatine gels[15], wheat gluten[16], and heat shrunk collagen[17,18] possess high elasticity. Chemical treatment may affect their crosslinkage, and so change their properties.

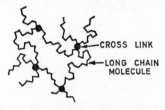

Figure 2.8 The ideal rubber network

The physical cause of high elasticity lies in the multiplicity of arrangements of the chain molecules. So its characteristics are totally different from ordinary elasticity.

3 Solids: Examples

3.1 SPAGHETTI

In the manufacture of spaghetti a stiff dough is mixed from semolina milled from durum wheat, and water. This dough is extruded through dies to yield long, circular, uniform strands which are then dried and referred to as spaghetti. They are about 500 mm long and 1·5 mm in diameter. Before rheological testing, the spaghetti rods are usually conditioned by placing them into a temperature controlled room at 25°C and 65 per cent RH for 48 hours.

Young's modulus is determined by placing a piece of spaghetti across two horizontally clamped supports and loading it as described in Chapter 2.2. Young's modulus was found to be $0·27 \times 10^{10}$ N/m².

The shear modulus as determined by the formula given in Chapter 2.2 was found to be $0·11 \times 10^{10}$ N/m².

The breaking strength has been determined by means of a Houns-field Tensometer. In this instrument the spaghetti rods are clamped between channelled metal jaws and are then stretched by turning a handle manually. The force applied is recorded by a mercury gauge. From the maximum force before rupture, the stress may be calculated by dividing the reading by the cross sectional area of the spaghetti piece. This is determined from four micrometer readings taken at right angles to each other near the point of fracture. Spaghetti rods extend only a little before breaking, so the cross sectional area changes little during stretching. The breaking strength was found to be $17·95 \times 10^6$ N/m².

If the material is isotropic, Poisson's ratio can be calculated using the formula given in Chapter 2.1 and the values for Young's modulus and shear (rigidity) modulus.

Whether or not the material is isotropic is determined by estimating the swelling ratios of spaghetti in water. Figure 3.1 shows a plot of the percentage swelling of length and diameter against time. Water penetrates slowly into the spaghetti and up to 80 min a distinct hard core is visible surrounded by doughy material. After 110 min this core is no longer visible, although still discernible by rubbing the spaghetti, when the hard core remains. The core disappears after 170 min. There is a steady rate of increase in swelling

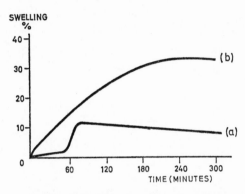

Figure 3.1 Swelling of spaghetti in water
(*a*) Length. (*b*) Diameter

diameter which falls off slightly with time. The swelling diameter is nearly steady at 33 per cent after 240 min. The longitudinal swelling is very slight at first because the solid spaghetti core does not allow extension. When this core has softened sufficiently, swelling is rapid to about 9 per cent after 90 min. This stage is followed by a slight decrease in length although the swelling diameter continues to increase for some time.

Since maximum swelling is 33 per cent in diameter and 9 per cent in length, this represents a swelling ratio of about 4:1.

It is apparent that spaghetti rods are not isotropic and hence the value for Poisson's ratio cannot be determined from the formulae given in Chapter 2.1.

Torsional and bending strength of macaroni have been investigated by Karacsonyi and Borsos[19]. The former value was found to be of the order of $2\text{--}7 \times 10^6$ N/m^2, the latter of the order of $10\text{--}30 \times 10^6$ N/m^2. Particularly the torsional strength was affected by hidden cracks in the macaroni and it was found that these flaws caused the macaroni to disintegrate during cooking. Hence there was a relationship between torsional strength and cooking properties.

3.2 EGGSHELL

A survey in 1953 showed[20] that approximately 9 per cent of eggs are cracked between the farm and the wholesaler. These eggs are not always a total loss, but often bacterial infection takes place through the crack. Salmonella (many of this group of bacteria cause food poisoning) was rarely found in sound eggs, but up to 3 per cent of cracked or leaking eggs were infected. Figure 3.2 shows an apparatus

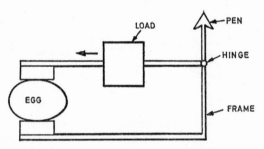

Figure 3.2 Apparatus for measuring the strength of egg shell

which has been used for measuring the strength of egg shell[21]. The egg is placed between two polished brass plates and the top plate loaded by a weight which is moved along the horizontal arm by means of a constant speed motor. In this way the load is increased by 100 g/s. The movement of the top arm is magnified 120 times and a record of compression is taken on a rotating drum.

It was found that the relation between load and deformation was nearly linear (Hookean). For the same species of bird the strength was related to shell thickness although the relationship was not close. There was no relation between shell strength and egg size. The harder the shell (measured by penetrometer) the stronger was the shell. When 365 hens' eggs were tested, it was found that the egg weight varied between 45 and 65 g. Mean shell thickness was

0·329 ± 0·002 mm, deformation before cracking 0·167 ± 0·0015 mm, and the load to break the shell 3·66 ± 0·042 kg.

When the eggs were cracked between parallel brass plates, it was found that the microscopic irregularities of the shell were deformed plastically during the test. When the shell was blackened and the maximum area of contact between shell and brass plate was measured with the aid of a microscope it was found to be 0·2–0·5 mm². From the value of breaking load divided by area of contact it has been suggested[21] that the breaking strength of the egg is as high as 12·2 tons/in² (19×10^4 N/m²). This argument is fallacious because the area of contact has nothing to do with the unit area of shell on which the load acts. The test is entirely empirical.

Impact tests have been conducted by dropping a steel ball on to the equator of the egg[22]. It was found that thinner shells are relatively stronger than thicker ones. The strength of egg shell has also been measured by emptying the egg, filling it with water and injecting it, using a hypodermic syringe fitted with a manometer. In this way the hydrostatic pressure at failure could be determined[23].

3.3 EMPIRICAL TESTING AND UNIVERSAL TEST SYSTEMS

Many important solid foods (meat, fish, fruit, and vegetables) are structurally very complex. They are almost always anisotropic. Hookean behaviour is only exhibited over a very small range of stress, if at all. There are almost always two or more phases and several component parts. Soft pears may contain gritty stone cells, peas with a soft interior have a much tougher seed coat, while French beans, asparagus or meat may be stringy or fibrous.

With such complex materials, fundamental testing is at present hardly possible and a large number of empirical testing devices have been developed during the last 100 years.

Some of these have been fitted to universal testing machines. These are instruments where two members or arms move towards or away from one another, in a carefully controlled manner. Various sample cells may be placed between them. The instruments can be applied to fundamental testing but are usually used empirically because the sample is so complex that although load or force and deformation can be determined, stress and strain cannot. Many of the cells fitted to universal instruments had previously been described as separate testing devices. Figure 3.3 shows some examples of cells used.

The three best known universal testing instruments are the Wolodkewitch Bite Tenderometer[24], the L.E.E. Kramer shear press[25] and the Instron testing machine[26].

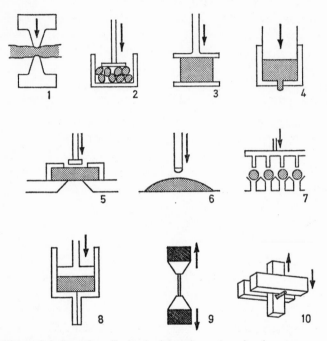

Figure 3.3 Sample cells used with universal testing instruments:
1. Meat tester   2. Gap extruder (grapes)   3. Plate compressimeter
4. Juice extractor (meat, etc.)   5. Shear strength tester   6. Apple penetrometer   7. Pea penetrometer   8. Capillary extruder
9. Extension tests   10. Meat shear unit

In 1955 a different instrument was developed by Proctor, incorporating a set of dentures. Resilient materials were used to simulate the tongue and lips[27]. From this instrument the Texturometer of General Foods has been developed which was first described[28] in 1963. The most important difference from Proctor's original instrument is that in the Texturometer the dentures are replaced by a plunger which moves up and down in a vessel fitted to a strain gauge. Each up and down movement simulates one chew. Figure 3.4 shows some tracings obtained with this machine. In each the right peak refers to the first

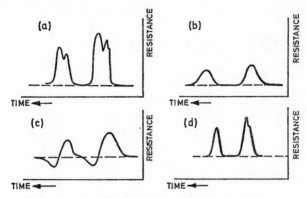

Figure 3.4 Tracings obtained with the Texturometer
(*a*) Dry dog food. (*b*) Cake crumb. (*c*) Pudding. (*d*) Bran flakes

chew, the left one to the second. The differences between the two peaks suggest a breakdown in the food. A descent below the base line is due to stickiness in the sample.

4 Newtonian Liquids: Characteristics and Measurement

4.1 NEWTONIAN VISCOSITY

What is viscosity? How can it be visualized? Imagine yourself stirring a cup of tea rather more vigorously than is consistent with good manners. Take the spoon out of the cup. The liquid rotation slows down and eventually stops. Why does the liquid rotation not go on for ever? It is friction within the liquid that slows it down. That liquid friction, the internal resistance of the liquid to flow, is called viscosity. It differs however, from solid friction in that for a true liquid, the smallest force (or traction) applied for a long time causes continuous flow.

Figure 4.1 shows a glass tube through which liquid flows at a slow rate. Through the thistle funnel a dye is introduced into the stream.

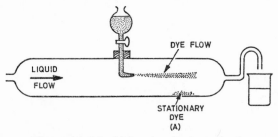

Figure 4.1 Liquid flow in a horizontal tube

When the liquid flow is arrested the dye, being heavier than the liquid, will collect in a small pool at the bottom of the tube. If the liquid is now allowed to flow again, it is found that the dye at A remains in a state of rest. The dye coming out of the jet at the thistle funnel moves in a straight line.

The liquid in the tube is, of course, continuous but one can imagine it to consist of very thin concentric layers. Under slow conditions of flow the centre layer will move fastest and the outermost layer will be stationary. The intermediate layers will show intermediate speeds of flow. There will be 'friction' between each imaginary layer. Figure 4.2 shows this arrangement of imaginary, infinitely thin, concentric layers.

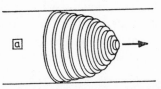

Figure 4.2 Streamline (telescopic) flow: Slow flow in a tube visualized as moving concentric layers

This type of flow is described as streamline flow. It has been defined as: a regular flow which can be described as the flow of a number of infinitely thin layers, each moving as an entity relative to the other.

Let us now imagine that we could remove a square from the layer closest to the wall and its neighbour ('a' in Figure 4.2) and magnify them enormously so that each layer is 1 m² in area and 1 m away from the other. If we apply a shear force of 1 newton to the surface of the upper layer (*see* Figure 4.3) and obtain a velocity of 1 m/s then the

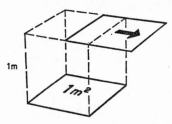

Figure 4.3 Definition of coefficient of viscosity

coefficient of viscosity, η, of the liquid equals one poiseuille (i.e. 1 Ns/m²). The poiseuille, Pl, is the unit of viscosity. In the older literature the poise is used as the unit of viscosity. One poiseuille equals 10 poise.

So the coefficient of viscosity, η, is a quantitative term, a numerical quantity. 'Viscosity' on the other hand, is qualitative. η is defined as the ratio of shear stress τ, to the rate of shear D, i.e.

$$\eta = \frac{\tau}{D} \quad \text{or} \quad \tau = \eta D$$

If η is a constant irrespective of the value of D, this equation is that of a Newtonian liquid.

A Newtonian liquid is a liquid for which a graph of shear stress against rate of shear is a straight line. The proportionality constant is called the coefficient of viscosity. The shorthand symbol is 'N' (*see* Figure 4.4), and the model a dash pot which is a convenient way of

MODEL	SHORTHAND SYMBOL	EQUATION
	N	$\tau = \eta D$

Figure 4.4 The Newtonian liquid

visualizing an ideal liquid. It is a container filled with liquid through which a plunger can be pushed up and down.

The Newtonian liquid is called after Sir Isaac Newton who first defined viscous flow. 'The resistance', he said, 'which arises from the absence of slipperiness in a fluid, is proportional to the velocity (other things being equal) by which the parts of the fluid are being separated from each other'[29].

The Newtonian liquid has no elastic properties, but is incompressible, isotropic, structureless, and, like the Hooke solid, does not exist in reality.

Nevertheless, many real liquids exhibit Newtonian behaviour over a wide and important range of shear stress. These are the liquids which the rheologist loosely refers to as 'Newtonian liquids'.

Two commonly used graphs showing the flow properties of Newtonian liquids are the plot of shear stress, τ, against rate of shear, D, or the coefficient of viscosity, η, against rate of shear, D. The shear rate is expressed in reciprocal seconds (s^{-1}). From Figure 4.5(a) it

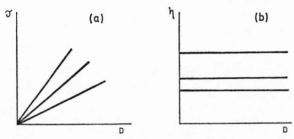

Figure 4.5 (a) Plot of shear stress τ against rate of shear D
(b) Plot of coefficient of viscosity η against rate of shear D

is apparent that the slope of the lines are given by η, from Figure 4.5(b) that η is a constant. To determine whether a liquid is indeed Newtonian, a doubling of the shear stress, τ, should result in a doubling of the rate of shear, D. If this is not the case the liquid is not Newtonian.

The coefficient of viscosity is very dependent on the temperature (*see* Chapter 4.4) and for this reason the values in Table 4.1 are given at 20°C.

TABLE 4.1

Some coefficients of viscosity (Pl) at 20°C

Hydrogen	87.6×10^{-5}	Ethanol	1.200×10^{-1}
Carbon dioxide	148.0×10^{-5}	Cottonseed oil	70.4×10^{-1}
Air	183.0×10^{-5}	Olive oil	84.0×10^{-1}
Chloroform	0.58×10^{-1}	Castor oil	986.0×10^{-1}
Carbon		Glycerol	149.0×10^{-1}
tetrachloride	0.969×10^{-1}		

The coefficient of viscosity is often referred to as the 'dynamic viscosity' to distinguish it from the 'kinematic viscosity' v. The latter is the viscosity coefficient that is directly measured in U-tube capillary

TABLE 4.2

Some high viscosity materials (Pl)

	η Poiseuille	Temperature
Sucrose melt	$1 \cdot 9 \times 10^4$	$124 \cdot 6°C$
Pitch	$51 \cdot 0 \times 10^9$	$0°C$
Soda glass	$11 \cdot 0 \times 10^{11}$	$575 \cdot 0°C$

viscometers where the shear stress is caused by the head of the liquid. This depends not only on the height of the column, but also on the density ρ, of the liquid. Also in some engineering calculations the density and dynamic viscosity terms occur together. To convert kinematic viscosity into the dynamic viscosity, the latter has to be divided by the density of the liquid.

Occasionally the term 'coefficient of fluidity', ϕ, is encountered in the literature. This is the reciprocal of the dynamic viscosity. The relationship is summarized in the equation:

$$\eta = v\rho = 1/\phi$$

4.2 STRUCTURE

Generally, simple liquids and true solutions are Newtonian. They have no connected structure and no measurable rigidity. Some suspensions or solutions of macro-molecules are Newtonian, but not many. As a rule of thumb, if the macro-molecular chain in the liquid consists of less than 1000 atoms, Newtonian flow properties are exhibited. However, the concentration is important. At a low concentration even large molecules may show Newtonian behaviour in solution. Generally, in order to show Newtonian behaviour, macro-molecular suspensions must have a loose and discontinuous structure. The particles must behave rigidly under the conditions of the test. Asymmetrical (rod or plate shaped particles) may even show some non-Newtonian effects at very great dilution. There must either be no interaction between them, or the interaction must be independent of the flow rate. The shear will occur in the solvent which fills the space between the particles.

4.3 TURBULENCE

If the liquid flow shown in Figure 4.1 is increased beyond a certain rate, the stream of dye breaks up into eddies. This condition is referred to as turbulence. Using water in the apparatus shown (tube diameter 36 mm, tube length 570 mm) it occurs at a flow rate of approximately 0·240 l/min and shows itself in the formation of eddies and vortices. It is a condition of Newtonian flow that it can only be observed below the point at which turbulence occurs. In other words, there must be a streamline flow. The change from streamline flow to turbulent flow is very sudden. Figure 4.6 shows the τ–D diagram for Newtonian and turbulent flow.

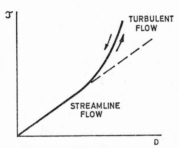

Figure 4.6 Shear stress, τ, plotted against rate of shear, D, in Newtonian and turbulent flow

Under turbulent conditions the curve is similar to that of a dilatant material (*see* Chapter 6.2). The steep rise in the curve is not due to the composition of the liquid and its viscosity, but to additional internal friction caused by the turbulent vortices. Under turbulent conditions, the flow behaviour of Newtonian and non-Newtonian liquids (*see* Chapter 6) tends to become indistinguishable. The first investigator[72] to study turbulence scientifically was Osborne Reynolds (1842–1912). In his honour the number which indicates the onset of turbulence has been named the 'critical Reynolds number' R_c. For flow through tubes:

$$R_c = \frac{2\rho Q}{\pi r \eta}$$

where ρ is the density of the liquid (kg/m³), Q is the volume (m³/s) delivered at the end of the tube, r is its radius (m), and η is the coefficient of viscosity of the liquid in Pl (Ns/m²). If the units of measurement are consistently chosen, R_c like Poisson's ratio, is dimensionless.

For a Newtonian liquid, R_c for flow through tubes is of the order of 2300. Pseudoplastic characteristics (*see* Chapter 6.2) tend to stabilize flow, and turbulence occurs at a higher R_c. If the coefficient of viscosity is extremely high, no turbulence may occur at all, because it may be experimentally impossible to obtain sufficiently high flow rates. For viscoelastic liquids (*see* Chapter 10) much lower R_c values are often found. The elongated particles which are often contained in such solutions act as nuclei for turbulence. Generally R_c decreases with increasing size and concentration of macro-molecules. With non-Newtonian liquids (*see* Chapter 6) R_c cannot be calculated because the coefficient of viscosity is not constant.

Viscosity measurements must take place below the critical Reynolds number, and no turbulence must occur. Concentric cylinder visco-meters give high stability of streamline flow particularly if the outer cylinder rotates rather than the inner one. A check should always be made and the R_c may be calculated or if suspended particles are present may be observed visually. The curves in Figure 4.6 are readily determined with a concentric cylinder viscometer. The point of divergence of the experimental curve from the straight line may be determined and all values above it discarded. Care must be taken that temperature rises in the liquid at high shear rates do not affect the results.

The boundary layer occurring in streamline flow is a barrier to heat transfer in industrial practice. For this reason turbulence is some-times artificially induced. On pumping however, more energy is required to move the liquid under turbulent than under streamline conditions, because more energy is dissipated in forming vortices and eddies. In fact the critical Reynolds number is of great importance in engineering calculations. As the transition from streamline to turbulent flow is not gradual but sudden, different equations are required for dealing with fluid flow above and below R_c.

4.4 VISCOSITY AND TEMPERATURE

The viscosity of liquids decreases very considerably with increase in temperature. As a rule, the higher the coefficient of viscosity, the greater will be its change with temperature. The decrease of the viscosity of water, for instance, is of the order of 1–3·5 per cent per °C, depending on the temperature of measurement. For castor oil it is 8 per cent between 20°C–21°C, and for pitch at the same temperature interval, 30 per cent.

For viscosity tests on foods the temperature should be controlled within ± 0·5°C by immersing the viscometer in a water bath. The thermometer should be graduated into 0·1 or 0·2°C divisions and the bath thermostatically controlled. If this is not possible a large beaker of water may be brought to the correct temperature with a Bunsen burner and the liquid in it should be well stirred. Its temperature should be regulated before, during, and after the test so that the variation between subsequent tests is no larger than 0·5 per cent of the flow time.

Many attempts have been made to relate the coefficient of viscosity to temperature in a quantitative manner. Early work was based on the fact that with an increase in temperature the volume of the liquid also increased. Hence there were fewer molecules in the same volume of liquid. It is, however clear that the viscosity of a liquid is not a function of its molecular volume alone. Interaction between molecules of the liquid is also important and this varies with the chemical composition of the liquid.

If $\log \eta$ is plotted against the reciprocal of the absolute temperature $(1/T)$, for most Newtonian liquids the results give a straight line (water is an exception to this rule) so:

$$\log \eta = \frac{B}{T} + C$$

where B and C are constants for the particular liquid. Table 4.3 shows some values of B and C.

4.5 VISCOSITY MEASUREMENT

In Chapter 2, shear was briefly discussed. It was stated that the deformation caused by a shear stress could be expressed as an angle. Earlier in this chapter it was pointed out that it is convenient to visualize a fluid made up of very thin layers. In shear these thin layers are moved or twisted in relation to each other.

TABLE 4.3

Values of B and C for various liquids[30] (see text)

	B	C
Methyl alcohol	927	0·9123
Ethyl alcohol	1178	0·7467
Propyl alcohol	1595	−0·0305
Butyl alcohol	1742	−0·3285
Formic acid	1255	0·8920
Acetic acid	970	0·4398

This gives us four simple types of shear, shown in Figure 4.7. Viscosity measurements are mainly based on these four types of

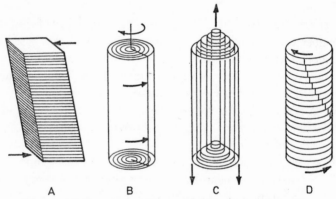

A   B   C   D

Figure 4.7 Four types of shear, visualized as: (*A*) Pushing over a deck of cards; (*B*) Twisting a nest of corkborers; (*C*) Pushing a nest of corkborers; (*D*) Twisting a stack of coins[31]

deformation. It should be noted that in B and D neither change in volume nor shape occurs. In all instances the measure of the deformation is α or L/H, the displacement gradient (*see* Chapter 2).

The methods of determining the viscosity of a liquid are all based on one concept: the motive 'force' equals the material constant (the viscosity) multiplied by some factor expressing the instrumental geometry (which is calculated or determined by calibration against a standard material), multiplied by some factor expressing the change or rate of change of shape of the material (this is measured).

There are basically five practical ways of determining the viscosity of a liquid. These are: (*a*) measuring liquid flow through tubes; (*b*) the fall of a solid body through the liquid; (*c*) rotational viscometry; (*d*) vibrational viscometry; and (*e*) using empirical techniques.

(*a*) *Flow through tubes*

Measurement is based on Poiseuille's Law. Poiseuille was a French physician (1797–1869) and the first to make exact measurements of viscous resistance. His law states that:

$$Q = \frac{\pi p r^4}{8 \eta l}$$

where Q is the volume of flow delivered through the capillary (m^3/s), p is the pressure difference between the ends of the capillary in N/m^2. This may be due to an externally applied pressure, or to the head of the liquid when it depends on the density of the liquid and its height. r is the tube radius (m), l is the tube length (m), and η is the coefficient of viscosity in Pl. It should be noted that the term 'capillary' in rheology means a narrow tube. It is much larger in diameter than the tubes for which this term is used in biology.

Determination of the absolute viscosity is rare because the instrument constants such as tube length and diameter would have to be known exactly. Usually one determines the relative viscosity using either water or sucrose solutions as reference liquids (*see* Appendix B). Water is the principal standard from which other values are derived. The viscosity of a liquid relative to that of a standard liquid, its 'relative viscosity' is found from the equation:

$$\frac{\eta_1}{\eta_2} = \frac{\rho_1}{\rho_2} \times \frac{t_1}{t_2}$$

η_1 and ρ_1 are the viscosity and density of the reference liquid and t_1 is the time for the flow of a particular volume of the liquid. η_2, ρ_2, and t_2 are the respective figures for the test liquid.

This relationship is only approximate because it ignores the kinetic energy correction. This is often necessary because the liquid emerges from the capillary tube with appreciable kinetic energy.

Hence not all of the pressure difference causing flow is used in overcoming viscous friction. The correction varies with the viscometer and the liquid. With liquid foods the kinetic energy correction is often neglected as it frequently falls within the allowed error of the determination[32].

The usual laboratory type of capillary U-tube viscometer is the Ostwald instrument which is available with capillary tubes of various diameters and may be used to measure[33] coefficients of viscosity in the range of 0·001–0·35 Pl.

Many modifications of the Ostwald viscometer exist and are described in the various laboratory catalogues. Figure 4.8 shows a

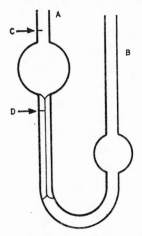

Figure 4.8 The Ostwald Viscometer

typical Ostwald viscometer and in Appendix A an experiment using this instrument is described.

(b) Fall of solid body through liquid

The measurement of the coefficient of viscosity by means of the fall of a sphere through an infinite volume of fluid is based on Stokes' Law (Sir George Stokes, 1818–1903, Irish physicist).

$$\frac{4}{3}\pi R^3(\rho_s - \rho)g = 6\pi\eta RV$$

$$\text{or } \frac{4}{3}R^2(\rho_s - \rho)g = 6\eta V$$

where R is the sphere radius (m), ρ_s is the density of the sphere (kg/m³), ρ is the liquid density (kg/m³), g is the acceleration due to gravity (9·81 m/s²), η is the coefficient of viscosity in Pl, and V is the velocity of the sphere (m/s).

The relative viscosity is determined by the equation:

$$\frac{\eta_1}{\eta_2} = \frac{(\rho_s - \rho_1)}{(\rho_s - \rho_2)} \times \frac{t_1}{t_2}$$

where t_1 and t_2 are the times (s) for the sphere to fall the same distance through the two liquids.

Figure 4.9 shows a suitable apparatus which is easily made from a measuring cylinder. The diameter and distance of the reference

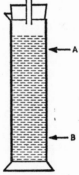

Figure 4.9 The Stokes' method for viscosity determination

marker from the ends must be very large compared with the ball diameter. In Appendix A an experiment using the apparatus is given.

This method is suitable for liquids of a viscosity higher than those normally used with U-tube viscometers ($1-3 \times 10^5$ Pl). The sphere is commonly a steel ball bearing of 1·5 mm diameter. For more viscous liquids a platinum ball may be used because platinum is denser than steel. If the liquid is opaque, a metal detector can be employed.

Instead of a falling ball, a rising sphere made of glass or wood can also be used.

(c) *Rotational viscometers*

There are three types of rotational viscometers, those with an external rotating cylinder, those with an internal rotating cylinder, and cone and plate viscometers.

An example of an instrument where the external cylinder rotates is the coaxial cylinder viscometer of the Couette–Hatcheck type. The coefficient of viscosity, η, is determined from the equation:

$$\eta = \frac{ked}{2\pi h \omega R}$$

where k is the torsional constant of the wire, e is the angular deflection of the inner cylinder, d is the gap between the cylinders, h is the height of the liquid between the cylinders, ω is the angular velocity of the outer rotating cylinder, and R is the mean radius of the cylinders. The equation is only valid if d is small compared with R.

For the same instrument:

$$\eta = \frac{Ke}{\omega h}$$

where K is the instrument constant. If the cup is always filled to the same level, h, and if the angular velocity of the outer cylinder, ω, is unchanged, both ω and h are part of K. Hence:

$$\eta = Ke$$

If a reference liquid of known viscosity, η, giving a cylinder deflection e_1 is available, then:

$$\frac{\eta_1}{\eta_2} = \frac{e_1}{e_2}$$

where η_2 and e_2 refer to the test liquid.

Commercial viscometers are usually standardized already by the manufacturer, and viscosity determinations become extremely simple: e_2 is determined and η_2 obtained directly from a chart or graph. Figure 4.10 gives a diagram of a rotating cylinder viscometer.

The inner cylinder is supported from a wire to which a mirror is attached. The twist of the wire is determined by shining a light on to the mirror and measuring the deflection of the light along a scale a little distant from the instrument.

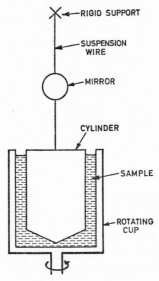

Figure 4.10 The rotating cylinder viscometer

End corrections   The viscosity equation above holds only for the narrow parallel gap between the two cylinders and not at the top and bottom. While in theoretical rheology rotating cylinder viscometers without bottoms are often of help, technical difficulties prevent their use in practice! Hence, end corrections have to be made. Several of these are possible. One can determine the end correction by filling the viscometer cup to different levels, or one can place mercury at the bottom of the cup. Often the viscometer is so designed that the inner cylinder is cone shaped rather than flat at the bottom, which greatly reduces the end correction. Other methods include the use of coaxial cone ends with the same rate of shear as in the cylinder annulus, the fitting of guard cylinders to shield the ends, and the trapping of an air bubble at the bottom of the cylinders.

It is important to note that on using the instrument there must be no slip at the wall, otherwise incorrect results are obtained.

Examples of coaxial cylinder viscometers with an internally rotating cylinder are those by Searle and Stormer. Here the inner cylinder is driven by a weighted string which runs over a horizontal pulley attached to the inner cylinder. The speed of rotation is inversely proportional to the viscosity of the liquid. A very effective, if expensive, type of instrument is the cone and plate viscometer. The earliest types were parallel plate instruments which measured viscous friction between parallel plates. Unfortunately the shear rate is then greater towards the circumference of the plates because the speed of rotation increases with the distance of a circle from its centre. (For the same rotation of a driving shaft, the edge of a large wheel runs faster than that of a small wheel.) This phenomenon was allowed for by increasing the gap at the edge and so the cone and plate viscometer was born.

(d) Vibrational viscometers

The oscillations of a sphere, a disc or a cylinder suspended on a wire are damped on immersion into a fluid. Oscillations can also be applied to the outer cylinder of a concentric cylinder viscometer. The damping depends on fluid viscosity, which may be determined in this way. The method is not often used and the equations are not simple because of the inertia of the systems[31,34,35].

(e) Compression of cylinders

This method may be used for the determination of the viscosity of fluids with very high viscosities (e.g. melted sugar) in the region of 1×10^5–10^8 Pl. The material is heated and a small cylinder is cast in a mould. Usually a tempering period is required during which the mould is kept at an elevated temperature for several hours. The material is then cooled, brought to the correct temperature and the coefficient of viscosity, η, in Pl, is then determined according to the equation:

$$\eta = \frac{\left(\dfrac{Wg}{3\pi r^2}\right)\left(\dfrac{1 - l/l_0}{2}\right) t}{2 \cdot 3 \log l_0/l}$$

where W is the weight (kg) used to compress the cylinder, g is the gravitational constant (9.81 m/s^2), π is 3.1416, r is the radius of the cylinder (m), l_0 is the original length of the cylinder (m), l is the final length of the cylinder (m), and t is the time (s).

(f) Empirical measurements

In empirical measurements no results in fundamental units (kg, m, s,) are obtained and test results depend entirely on the instrument used. If only Newtonian liquids are measured the instruments can, however, sometimes be calibrated. The results cannot be compared with those obtained with any other instrument, or a different technique. Nevertheless, such results may still be useful. It is possible to measure the torque on a stirrer shaft by attaching a strain gauge, or Watt meters or Watt-hour meters may be attached to the stirrer motor. In these examples, friction within the motor bearings or gears may affect the result and for this reason motors are often run for up to half an hour before readings are taken. The stirrers themselves may produce turbulence. A well known laboratory viscometer is the Brookfield rotating spindle instrument, which is immersed in a large 'sea' of liquid. The motor drives a verticle spindle and a paddle, a disc or a cylinder. The attachments are driven through a calibrated spring and the angular lag of the spindle behind the rotation of the motor is proportional to the coefficient of viscosity. Unfortunately the gap is infinitely large and for that reason the viscosity cannot be determined in absolute units unless the fluid is Newtonian.

Occasionally the rolling ball viscometer devised by Höppler is employed. This also will only give fundamental values of viscosity with Newtonian liquids. The instrument is extremely sensitive in the ranges of 10^{-3}–10^4 Pl. It is for instance possible to determine the viscosity difference between tap water and distilled water. The fall time of the sphere should be less than 17 mm/s or turbulence will occur. The ratio of the ball diameter to the diameter of the tube in the Höppler viscometer is of the order of 1.004–1.342, i.e. the ball nearly fills the tube.

Other empirical instruments are the cup, or 'orifice' viscometers of the Ford, Engler or Redwood type. Such an instrument is shown in Figure 4.11.

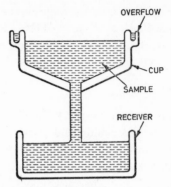

Figure 4.11 An empirical cup viscometer

The outlet is an orifice and not a tube and hence there is no uniform streamline flow. The pressure head also varies during the test. Much of the energy is used to provide kinetic energy so that the viscosity is only one factor affecting the result.

Penetrometers are at present also empirical instruments because the mathematical equations have not yet been developed. Penetrometers will be considered further when plastic materials are discussed (*see* Chapters 8 and 9).

(g) *Continuous measurement and automatic control*

Normally, viscosity is controlled in a factory by the collection of samples and subsequent viscosity determination in the laboratory. There is an inevitable time delay and for that reason continuous measurements have been developed. Almost all the types of viscometer described have been used for this purpose[31].

Rotating discs or cylinders may be immersed in a funnel in which the product rises and overflows continuously over the rim. With capillary viscometers part of the product may be diverted from the main stream and passed through a gear pump, delivering an accurate amount of the material through the capillary. The pressure difference at the ends of the capillary tube may then be measured. This is contrary to the normal practice where the pressure difference is kept steady and the delivered volume is measured.

With the falling ball principle, specially machined floats are suspended in an ascending stream of the product. The floats fall under gravity and obtain an equilibrium position which depends on the viscosity of the product. Again the flow rate must be strictly controlled. Sometimes a small piston is dropped through a column of liquid, only to be raised mechanically for the next determination step.

Oscillating techniques are also employed in which a thin metal strip is oscillated in the fluid. The fluid dampens the oscillations and these must be increased to maintain the original amplitude. The input is proportional to fluid viscosity.

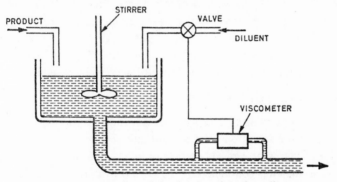

Figure 4.12 Automatic and continuous viscosity adjustment

Continuous viscometers can be used to operate diluent valves automatically and so adjust the viscosity with a smaller time delay than would be necessary if the material had to be removed to the laboratory for test. Figure 4.12 shows the layout of such an automatically operated process.

5 Newtonian Liquids: Examples

Examples of liquid foods exhibiting nearly Newtonian behaviour are carbonated beverages, alcoholic drinks, (as long as they do not contain long chain molecules), meat extracts, corn syrup, sucrose and salt solutions, maple syrup, some light cooking oils, some honeys, milk, dilute solutions of gum arabic, and water. The following three examples have been chosen to illustrate Newtonian behaviour.

5.1 SUGAR SOLUTIONS

Several workers have studied the viscosity of sucrose solutions, because they are often used to calibrate viscometers. They give a range of viscosity well above that of water. The most thorough of these studies was conducted by the U.S. Bureau of Standards which investigated the viscosity of sucrose solutions between 0–35°C and at concentrations of 20–75 per cent by weight. Table 5.1 shows the coefficient of viscosity of sucrose solutions at 20°C.

If log η is plotted against grams of sucrose in 100 g water, a line is obtained which is almost, but not quite, straight.

At the same total solids content, invert sugar solutions (invert sugar is a mixture of dextrose and lævulose) have a lower viscosity, and dextrose solutions a higher viscosity than those of sucrose (*see* Figure 5.1).

The viscosity of mixed sugar solutions is roughly additive and can be calculated approximately.

In the manufacture of sweets (candy) a saturated solution of sucrose in water is boiled. To avoid crystallization at elevated temperature, potassium hydrogen tartrate may be added to invert about one-third of the sucrose. As the water evaporates the temperature of the solution rises. When a sample is removed and cooled to room temperature it

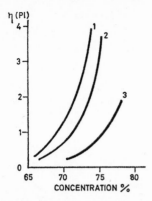

Figure 5.1 Coefficient of viscosity at 20°C of:
(1) dextrose; (2) sucrose; (3) invert sugar solutions[37]

TABLE 5.1

Coefficient of viscosity of sucrose solutions[36] at 20°C

%	Sucrose g/100 g water	η (Pl)	%	Sucrose g/100 g water	η (Pl)
20	25	0·0020	50	100	0·0155
25	33·3	0·0025	55	122·2	0·0283
30	42·9	0·0032	60	150	0·0589
35	53·8	0·0044	65	185·7	0·1482
40	66·7	0·0062	70	233·3	0·4850
45	81·8	0·0095	75	300	2·3440

exhibits very typical rheological properties and according to these, bears a name given by the sugar boiler. At a boiling point of 110°C the 'thread degree' is reached. The sugar is dense enough to form threads when a fork is inserted into it and pulled away. 115°C is referred to as 'feather degree'. The sugar will blow off a fork as feather-like fluff. At the 'ball degree' the sugar solution, once cooled to room temperature, allows a soft ball to be formed between the fingers. 'Crack degree' indicates that the ball is now hard and brittle (*see* Table 5.2). At present the values of viscosity given in Table 5.2 must be regarded as tentative, because it is not known how far Newtonian behaviour is retained at the high sugar concentrations.

TABLE 5.2

The relation[37] between technical nomenclature and boiling point (BP), total solids (TS),and coefficient of viscosity (η)

Technological description	BP°C	TS %	η (Pl)
Thread	110–113	75–80	1·8–10
Feather	115–119	87–88	approx. 4–100
Soft ball	121–124	90–91	approx. 10^5
Hard ball	127–130	92–93	2–3 × 10^5
Soft crack	135–138	95	approx. 14 × 10^8
Hard crack	144–149	97–98	?

It is apparent from Table 5.2 that the viscosity of boiled sugar solution, once brought to room temperature, may vary from less than 2 poiseuille to more than 100 million poiseuille. From 0·001 to 0·35 Pl, capillary viscometers are normally used. From 1–3 × 10^5 Pl, a falling ball of steel or platinum is employed. Between 10^5–10^8 Pl, the cylinder compression technique is used.

5.2 MILK

Milk composition

Milk is essentially an aqueous emulsion of butter fat globules of 0·0015–0·01 mm diameter. It contains about 87 per cent water, 4 per cent fat, 5 per cent sugar (mainly lactose), and 3 per cent protein (mainly casein). There are also many other constituents such as carotenoids, fat soluble vitamins, mineral salts, enzymes, various cells, and dissolved gases. Rheological investigations have involved mainly the relationship between viscosity on the one hand, and composition, heat treatment, homogenization, and non-Newtonian characteristics on the other. Viscosimetric measurement is sometimes difficult since cream may separate out during the test. Milk is very nearly Newtonian but shows a slight decrease in viscosity with rising shear stress. This is shown in Figure 5.2 which is taken from the work of Bateman and Sharp[38]. These workers used a U-tube viscometer fitted with an air pump reservoir and a manometer.

When the total solids content of the milk is increased, pronounced non-Newtonian behaviour may occur in both evaporated and skim milk. Aged, sweetened, condensed milk shows very complex visco-elastic properties which have been referred to as the Weissenberg

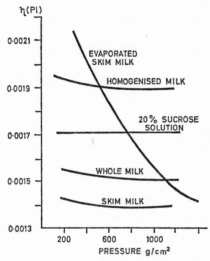

Figure 5.2 Decrease in the coefficient of viscosity of various types of milk with increasing shear force[38]

effect (*see* Chapter 10). The viscosity behaviour of cream is also varied and complicated. As the fat content is, however, reduced by the addition of water, anomalies tend to disappear and the rheological properties tend to become Newtonian again.

Viscosity and composition

Skim milk is less viscous than whole cream milk because the viscosity increases with fat content. The viscosity also increases with increase of non-fat solids but neither relationship is simple. Therefore the original hope of determining adulteration of milk with water by viscosity measurement was not fulfilled. In skim milk, additions of casein increase the viscosity considerably but the addition of as much as 5 per cent lactose has little effect. This is to be expected because relatively small molecules have little effect on viscosity as compared with large molecules. Table 5.3 shows the effect of dilution with water on the viscosity of fresh Guernsey milk.

TABLE 5.3

**Aqueous dilution and the coefficient of
viscosity of fresh Guernsey milk[38]**

Dilution % (by vol.)			η *Pl* (25°C)
0% water	100%	milk	0·001457
10	„	90	„	0·001381
20	„	80	„	0·001319
30	„	70	„	0·001258
50	„	50	„	0·001143
80	„	20	„	0·000993
100	„	0	„	0·000894

Viscosity and temperature

As with all liquids, the viscosity of milk decreases with increasing temperature. Figure 5.3 shows the relationship between temperature and viscosity for whole milk.

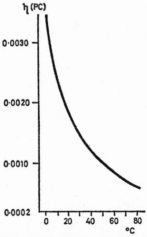

Figure 5.3 The relation between temperature and the
coefficient of viscosity of whole milk[39]

Several workers had determined the relation between temperature and viscosity of milk and obtained slightly different results. Cox[39] evaluated these statistically and developed the following empirical equation:

$$\eta'_\theta = \frac{\eta'_{20}}{1 + \alpha(\theta - 20) + \beta(\theta - 20)^2}$$

where η'_θ is the viscosity at the temperature of the experiment, η'_{20} the viscosity at 20°C, α is 0·00723, β is —0·000156, and θ is the temperature of the experiment in °C.

Milk is often heated for 15 s to 72°C to kill pathogenic bacteria and this increases the shelf life. This treatment is referred to as 'pasteurization'. It increases the viscosity of the milk slightly.

Viscosity and homogenization

Milk is often passed through a homogenizer to reduce the size of the fat globules. If these are too large, they tend to coalesce later and cream may separate out on standing. This is avoided by homogenization. As the globules are reduced in size their number, their specific surface, and the viscosity of the milk increase. Table 5.4 shows the relation between the homogenization pressure and viscosity increase.

TABLE 5.4

Relation between homogenization pressure and viscosity of milk[40]

Homogenization pressure 1000 lb/in²	Viscosity increase %
1	7·1
1·5	9·2
2	11·9
3	13·7
3·5	15·0

Viscosity and age

The viscosity of skim milk and homogenized or unhomogenized pasteurized milk increases with age. The reasons are unknown. It is certain that bacterial and enzymatic changes occur which result in a change of pH. The viscosity of milk is lowest at the natural pH and it is possible that the very slight increase in pH causes the increase in viscosity (*see* Table 5.5).

TABLE 5.5

Ageing at 4°–6°C of skim milk[38]

Age (days)	η (Pl) at 25°C	pH	Titrable acidity as % lactic acid
1	0·001433	6·78	0·14
3	0·001467	6·80	0·13
6	0·001521	6·81	0·13
9	0·001525	6·83	0·12
15	0·001551	6·84	0·11
21	0·001555	6·84	0·11

Zero gravity behaviour

Zero gravity conditions (weightlessness) may occur either during space flight or free fall in an aircraft during a carefully planned trajectory. The latter has been used frequently to simulate space flight conditions to study pumping of rocket fuels or the feeding of the astronauts.

Under weightless conditions the behaviour of liquids is governed by surface tension, surface adhesion, viscosity, and inertia. Generally, wetting liquids spread from their containers (or, in an enclosed vessel, adhere to the wall with a free space in the centre) as long as there are kinetic energy or thermal differences. Non-wetting liquids, by contrast tend to form spheres.

Some experiments with milk have been reported[71]. Under normal gravity, a stream of milk hitting a smooth metal plate will splatter. When milk is squirted on to the plate under zero gravity conditions, it will tend to form a globule on the surface without splattering (the 'non-splatter' phenomenon). When milk is squirted into a dish of milk at zero gravity it will rebound from the dish, and form a 'mushroom' shape. The top of the 'mushroom' tends to separate as a sphere and float in space.

When milk in a dish is struck with the underside of a spoon, it will normally splash upwards and out of the container. At zero gravity the spoon will only penetrate the surface and enter the milk without a splash (the 'non-splash' phenomenon). This is thought to be due to absence of fragmentation of the resultant wave which is normally due to gravitational shear forces. It is analogous to covering the surface of water with oil.

5.3 OILS

Oils are essentially esters of glycerol and long chain fatty acids. Usually the naturally occurring fatty acids have an even number of carbon atoms (16 or 18). Some are saturated like palmitic or stearic acid and some are unsaturated, i.e. they contain double bonds. Oleic has one, linoleic has two and linolenic has three double bonds. These double bonds are reative. During hydrogenation two hydrogen atoms are added to each double bond. This reaction causes an increase of the melting point and oils are thus converted into fats. If the double bonds from one fatty acid react with those of another, large polymerized structures may result. The relative number of double bonds in a fat or oil may be determined analytically by reacting with iodine (iodine value).

There is no strict dividing line between oil and fat. Oils are liquid at room temperature and fats are solid. For instance the fatty material extracted from wheat flour is referred to as 'flour fat' in the U.K. and as 'flour oil' in the U.S., a confirmation of the fact that the Americans generally over-heat their rooms!

The term 'oil' is also commonly applied to non-fatty products such as paraffin oils and essential oils. These are not considered here.

Structure

The molecules of oils join in groups of 'tuning forks', hence the structure is asymmetrical (*see* Figure 5.4). Oils are normally Newtonian but at very high shear rates there may be a curvature towards the D axis in the τ–D diagram (*see* Figure 6.2). (This behaviour will be referred to as pseudoplasticity in Chapter 6.) It may be due to the alignment of the unit cells at high shear stresses which would cause a decrease in internal friction.

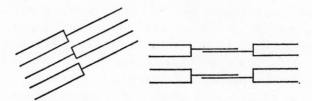

Figure 5.4 Molecular association of oil molecules

Effects of molecular size and shape

The bigger the molecule of an oil, the higher is its viscosity. Palmitic acid with sixteen carbon atoms has a lower viscosity than stearic acid which has eighteen. Completely hydrogenated cottonseed oil, which consists mainly of the glycerol ester of palmitic and stearic acid has a higher viscosity still (*see* Table 5.6).

TABLE 5.6

Coefficient of viscosity at 90°C (Pl)

Palmitic acid	0·005
Stearic acid	0·006
Hydrogenated cottonseed oil	0·012

All oils have a fairly high viscosity because of their long chain structure. The longer the chain of the fatty acids, the higher the viscosity. Polymerized oils (polymerization through carbon—carbon double bonds) have a very much higher viscosity than non polymerized oils.

The viscosity of an oil also increases with saturation of the carbon—carbon double bonds (*see* Table 5.7).

TABLE 5.7

Iodine value, melting point, and viscosity (50°C) of peanut and hardened peanut oils[41]

	Iodine value	Melting point	η (Pl)
Peanut oil (PO)	90·9	liq.	0·0241
Hardened PO	72·3	30°C	0·0294
Hardened PO	61·1	40°C	0·0311
Hardened PO	39·2	50°C	0·0330

Molecular interaction

The greater the molecular interaction, the greater the viscosity. The main constituent of castor oil is ricinoleic acid containing 18 carbon atoms with one hydroxyl group in the 12 position. Hydroxyl groups form hydrogen bonds and for this reason the viscosity of castor oil is higher than that of any other similar oil (*see* Table 5.8).

TABLE 5.8

Coefficient of viscosity (Pl) of various oils at 20°C

Cottonseed oil	0·070
Olive oil	0·084
Rapeseed oil	0·163
Castor oil	0·986

In free fatty acids bonding occurs through the carboxyl (COOH) group. Hence the esters have a lower viscosity than that of the acid itself, because here the carboxyl group becomes blocked. However, once esterified, the viscosity increases with increasing chain length of the alkyl radical (*see* Table 5.9.)

TABLE 5.9

Coefficient of viscosity (Pl) of oleic acid and oleic acid esters at 30°C

Oleic acid	0·0230
Methyl oleate	0·0049
Ethyl oleate	0·0052
Propyl oleate	0·0059

5.4 THE GLASSY STATE

The arrangement of molecules in a liquid is largely random. On solidification, the molecules usually arrange themselves in a crystalline pattern (e.g. metals, ice). Sometimes, when the temperature is dropped very rapidly, the liquid becomes so viscous that the molecules are unable to move into orderly patterns quickly enough and the material solidifies with random orientation. A typical example is window glass and for that reason the phenomenon is generally called 'the glassy state'. A liquid or supercooled liquid is said to be in the glassy state when its viscosity is more than 10^{12} Pl. It supports its own weight and its breaking strain is of the order of 0·04–0·05, i.e. it is brittle. The ultimate test of the glassy state is by X-ray diffraction. Here, this shows very small regions of orderly molecular arrangement, while in crystals there are large regions and in gases none. Examples of materials that can be found in the glassy state are alcohols, fats, and very concentrated sugar solutions. The glassy state is often of importance in food products[42]. Boiled sweets are in the glassy state and appear quite clear. When they are placed into a moist atmosphere, a layer of water is adsorbed on their surface and as a result the viscosity in the surface layer decreases rapidly. The sugar molecules are now free to move into their preferred position and crystallization may occur.

If crystallization occurs, water is liberated and the viscosity on the surface of the glassy mass falls further. Hence a chain reaction is set up and crystallization proceeds from the outside to the inside of the sweet. As a result it becomes opaque and sticky. Development of stickiness is typical during crystallization of glassy sugars. It may cause lumping and caking in milk powder and freeze-dried products. Usually the glasses of mixed sugars are less stable than those of single sugars, but glucose added to sucrose at about 20 per cent by weight causes greater stability. For this reason 'confectioners glucose' (hydrolysed starch) is often added to boiled sweets but this increases their cost. Crystalline lactose may cause grittiness in ice cream. In freeze-drying development of sugars in the glassy state is usually avoided, because it retards drying by occluding water.

6 Non-Newtonian Liquids : Characteristics and Measurement

6.1 APPARENT VISCOSITY

For a Newtonian liquid (*see* Chapter 4) the flow behaviour is given by the equation:

$$\tau = \eta D$$

This means that the shear stress is equal to the resultant rate of shear multiplied by a constant. The constant is the coefficient of viscosity. The τ–D graph is a straight line and passes through the origin. The slope of the line is given by η. Since η is a constant, a single determination can completely characterize the flow behaviour of the liquid (*see* Figure 6.1(*a*)).

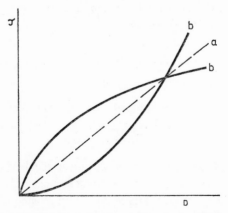

Figure 6.1 Newtonian (*a*) and non-Newtonian (*b*) flow curves

There are very many liquids employed in the food industry for which this simple relationship does not hold. Such liquids are often suspensions of solids or emulsions of liquids in a liquid medium. Both can be called dispersions. The discrete particles (the 'discontinuous phase') may interact with each other and also with the medium (the 'continuous phase') by which they are surrounded. If the interaction depends on the flow rate then the coefficient of viscosity is no longer a constant. Figure 6.1(*b*) shows, that for such systems a one point measurement is no longer adequate because it does not characterize the flow curve. How can such behaviour be recorded?

Instead of the coefficient of viscosity, η, an 'apparent' coefficient of viscosity, η_{app}, is often used. The latter is defined like the former, as shear stress divided by the rate of shear:

$$\eta_{app} = \frac{\tau}{D}$$

The apparent viscosity is no longer a constant, it depends on the shear stress and a graph will show the shear stress associated with its value of shear rate, (either as a flow curve of τ against D or η_{app} against D). Such liquids are called non-Newtonian, provided they also show continuous flow even for the smallest applied force.

A non-Newtonian liquid is defined as a liquid exhibiting uniform flow, but where the relation between shear stress and rate of shear is not constant. (The viscosity is not constant.) There are broadly four types of behaviour. These are:

	Time independent (steady state)	Time dependent
Thinning	Pseudoplasticity	Thixotropy
Thickening	Dilatancy	Rheopexy

These terms are the result of historical accident and have no rheological significance beyond the type of flow curve. Dilatant liquids do not necessarily dilate and there is nothing 'pseudo' about pseudoplasticity. These four types of flow behaviour will now be considered.

6.2 TIME INDEPENDENT (STEADY STATE) FLOW
(a) *Pseudoplasticity*

While subjected to high rates of shear (e.g. stirring) the liquid is thinner than when sheared slowly. The apparent viscosity depends on the rate of shearing but not on the length of time that shearing has proceeded. The τ–D curve is not a straight line. The shear rate increases more rapidly in proportion to the shear stress so the apparent viscosity decreases with increasing shear rate. For every τ there is a definite D and vice versa. Hence, the phenomenon is referred to as showing 'steady state' behaviour. The two types of flow curve τ against D and η_{app} against D are given in Figure 6.2. The arrows

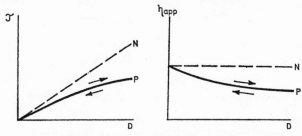

Figure 6.2 Pseudoplastic flow curves

indicate increase or decrease of shear rate D in a continuous experiment, for instance in a concentric cylinder viscometer. The name 'pseudoplasticity' derives from extreme examples in which rates of shear at low stresses are very small so that the graph lookes rather like that of the Bingham material (*see* Chapter 8).

(b) *Dilatancy*

This is a time independent thickening at high rates of shear, the opposite of pseudoplasticity. The curve is not linear but values of τ and D are uniquely related (steady state thickening). η_{app} increases with increasing D. It is the opposite of pseudoplasticity (*see* Figure 6.3).

6.3 TIME DEPENDENT FLOW
(a) *Thixotropy*

Thixotropy is time dependent softening. The plot is curved and similar to pseudoplasticity, in that η_{app} decreases with increasing shear rate. It differs however, in that the decrease is not only related to shear rate but also to time. With the same shear rate the η_{app} decreases with time so η_{app} depends not only on the rate but also on the

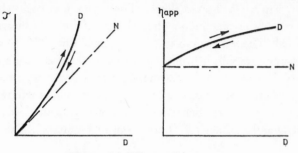

Figure 6.3 Dilatant flow curves

duration of shear. The τ–D curve is plotted by increasing D by definite intervals to a maximum and then decreasing it in the same way to zero. The time intervals between the readings and the readings themselves must be kept constant (*see* Figure 6.4). When the rate of

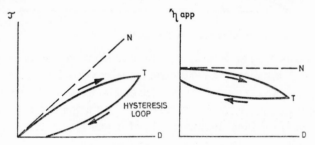

Figure 6.4 Thixotropic flow curves

shear is decreased the material only slowly thickens to give its original behaviour.

If the τ–D diagram is time dependent, non-Newtonian flow results in a loop (*see* Figures 6.4 & 6.5). This is referred to as a hysteresis loop. (The term hysteresis is also applied to delayed elastic phenomena, hence a general definition of hysteresis is: a deformation process where the loading and unloading parts do not coincide, but form a loop.)

An important consequence of this time dependency is that the testing routine must be strictly timed. Not only the test itself but the previous history of the sample (handling) must be controlled. For instance, sampling and filling the viscometer cup must be subjected to a definite routine. This is not necessary with time independent samples. The material is said to have a rheological memory.

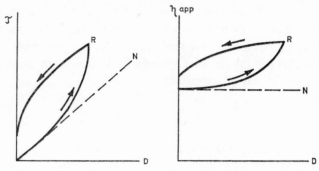

Figure 6.5 Rheopectic flow curves

(b) Rheopexy

This is time dependent thickening. The τ–D relationship is curvilinear and τ and D are not uniquely related. There are two values of D for each τ and vice versa if the cyclic testing routine is followed. Rheopexy is the opposite of thixotropy and the same time-control on testing applies (*see* Figure 6.5).

6.4 MEASUREMENT OF APPARENT VISCOSITY

Two aspects must be considered. Firstly special conditions of test are necessary because the viscosity is dependent on shear rate D. Secondly, special conditions arise through the viscosity being dependent not only on rate but also on duration of shear. The first applies to all non-Newtonian liquids. The second only to those that are time dependent (thixotropy, rheopexy). Obviously, instruments which give empirical results with Newtonian liquids, also give empirical results with non-Newtonian liquids. These are orifice viscometers, penetrometers, and instruments working under turbulent conditions.

The falling ball instrument cannot be used with non-Newtonian liquids to obtain results in fundamental units because the mathematical problems have not yet been solved. U-tube viscometers with a variable pressure head cannot be used because the shear rate varies. For capillary tubes with constant pressure heads the situation is slightly different. As is apparent from Figure 4.2, the shear rate in a capillary tube varies from zero at the tube axis to a maximum near the wall, so the apparent viscosity varies within the limits determined by the non-Newtonian properties of the sample itself. For time independent liquids Reiner[3] has made an attempt at a correction.

It is also possible to solve systems for flow through tubes using quantities $P*$ and $Q*$ defined by:

$$P* = \frac{pr}{2l} \quad \text{and} \quad Q* = \frac{4Q}{\pi r^3}$$

and then finding the D corresponding to $\tau = P*$.

This value of D equals

$$\tfrac{1}{4}\left(3Q* + P\frac{\mathrm{d}Q*}{\mathrm{d}P*}\right)$$

where $\mathrm{d}Q*/\mathrm{d}P*$ is the slope of the tangent of the $P*$, $Q*$ graph for the value of $P*$.[46,47]

It should be noted that this argument only applies if the liquid is not time dependent.

For time dependent liquids the capillary tube cannot be used because apart from the radial shear variation, there is no control over the duration of shear and no hysteresis curve is obtainable.

Viscometers conveniently used for non-Newtonian liquids are the narrow gap concentric cylinder and the cone and plate viscometers. Here the shear rate is the same throughout the sample provided any end effects are minimized. Shear rate can be varied in a controlled manner by varying the speed of rotation and the time taken for each speed. The experimental procedure is critical. Speed adjustment (increase or decrease) and the reading of corresponding values must be made without stopping the rotation. The speed is first increased in regular steps and then decreased in regular steps or if recording is employed, a smooth cycle of change can be used. Care must be taken in filling the instrument particularly if the material is time dependent (rheological memory). There must be no slip at the wall and if the material is viscoelastic and tends to climb up the inner rotating cylinder or bulges out of the gap in the cone and plate viscometer the experiment must be discontinued (*see* Weissenberg effect, Chapter 10).

6.5 STRUCTURE OF NON-NEWTONIAN LIQUIDS

Simple liquids and true solutions are usually Newtonian. Non-Newtonian liquids are usually very complex and consist of more than one phase, although polymer solutions may be treated as a single phase. There is always a continuous phase and one or more dispersed phases. In spite of much work the relation between rheology and structure of non-Newtonian liquids is usually quite obscure.

Qualitatively, the rheology of a dispersed system depends on the properties of the continuous phase, the dispersed phase, and the interaction between the two. In the continuous phase, viscosity, chemical composition, pH, and electrolyte concentration are of importance. In the dispersed phase, which may be liquid or solid, i.e. emulsion and suspension respectively, volume concentration (percentage of one in the other), the viscosity (if an emulsion), particle size, shape, size distribution, and chemical composition play a role. The interaction between the two phases may be affected additionally by stabilizing and surface active agents, and the properties of any stabilizing film may modify the behaviour.

Perhaps one can visualize a thinning system as a fluid where either chemical bonds are broken or where the particles align with increasing shear rate. Thus viscous friction decreases with increasing shear rate. In time dependence the *status quo* is reached slowly; in steady state phenomena, rapidly. With thickening systems electrical forces have been suggested which are stronger at higher shear rates[31,35].

6.6 POWER LAWS

In pseudoplastic and dilatant systems the τ–D relationship is not a straight line but unique. For each τ there is one corresponding value of D only. For this reason the curve can sometimes be described mathematically:

$$\tau = KD^n$$

where K and n are constants. K has been called the 'consistency index' and n the 'flow behaviour index'. The latter is a measure of departure from Newtonian behaviour. If $n = 1$, the material is Newtonian and $K =$ the coefficient of viscosity. If n is larger than 1, thickening occurs and the material is dilatant, if n is smaller than 1, thinning occurs and the material is pseudoplastic (*see* Figure 6.6).

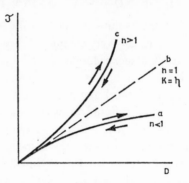

Figure 6.6 Pseudoplastic (*a*), Newtonian (*b*) and dilatant flow (*c*) '*n*' is the flow behaviour index

It is important to note that the power law only applies over the measured range. The equation must not be used to extrapolate the curve. K and n have no physical basis but are purely the mathematical description of an experimental curve. Nevertheless they are valuable

TABLE 6.1

Viscosity data for some fruit and vegetable products (TS = Total Solids)

			Temp.	n	K	ref.
Soups and sauces			12·8°C	0·51	3·6–5·6	43
Apple sauce			24	0·65	0·5	44
Banana puree			24	0·46	6·5	44
Tomato juice	5·8% TS		32	0·59	0·22	45
,,	,,	12·8% ,,	32	0·43	2·0	45
,,	,,	16·0% ,,	32	0·45	3·16	45
,,	,,	25·0% ,,	32	0·41	12·9	45
,,	,,	30·0% ,,	32	0·40	18·7	45

in practice. The power law should not be used if τ and D are not uniquely related, because for each value of τ there are many values of D. (Two values in a rigidly controlled instrumental cycle.) There is obviously no justification for a power law if the material described is in an unknown intermediate state of thixotropic breakdown or rheopexy. The power law should only be used if time independence of the material has been established by previous experiment. Table 6.1 shows some values of K and n.

7 Non-Newtonian Liquids: Examples

Almost all liquid foods except those listed at the beginning of Chapter 5 have non-Newtonian properties.

In spite of this, really good, fundamental rheological investigations on such complex systems are scarce.

Often, it is not stated which viscometer has been used and most of those employed are not suitable for non-Newtonian liquids. Occasionally, turbulence is deliberately produced during measurement to keep the dispersed phase in suspension but no fundamental measurement can be taken under turbulent conditions. Most commonly, only the upcurve of the τ–D plot is presented. This is then extrapolated by means of a power law. This is unjustifiable. Before applying a power law, check must be made to see that the τ–D plot is uniquely related, so that for each τ there is only one value of D. Until this is confirmed both the up and down curves must be determined. Under no circumstances must the curve be extrapolated (*see* Chapter 6).

Empirical rheological tests are often very useful, but faulty methods must be rejected. Recently, the following experiment was described in the literature: the viscosity of tomato juice was required and because the pulp tended to settle a magnetic stirrer was used to keep it in suspension. A Brookfield viscometer with a disc spindle was then inserted and the coefficient of viscosity in Poise was directly determined from the dial and reported. The result is, of course, meaningless and the reader will have no difficulty in spotting the three errors that have been made.

To illustrate non-Newtonian behaviour, honey, soup and hydrocolloid solutions will be given as examples.

7.1 HONEY

Honey is a concentrated solution of sugars with other minor constituents, prepared by bees from the sweet juices of plants. Nectar is exuded by the flowers and collected by bees which add the enzyme invertase. This splits the sucrose contained in the nectar into dextrose and fructose. The material is then transferred to the honey combs and loses water.

Nectar			*Honey*	
Sucrose	20%		Dextrose	35%
		Invertase	Fructose	40%
		$\xrightarrow{}$		
Water	80%	Water loss	Water	15%
			Undetermined material	10%

The undetermined material comprises about 2 per cent sucrose as well as protein, dextrans, organic acids, essential oils, vitamins, minerals, pollen grains, yeast, bacteria etc. Although there is a relatively small amount, this 'undetermined material' has a large effect on the viscosity behaviour of honey. Many honeys are, like sugar solutions, Newtonian, as for instance clover honey. Its viscosity depends on the percentage of water and the temperature. In one study[49] on various clover honeys, the coefficient of viscosity at 20°C varied from about 10 to about 50 Pl.

Not all honeys are, however, Newtonian. It has been known for a very long time that heather honey sets to a gel in the combs and is difficult to extract by the usual process of centrifugation. On stirring, its viscosity decreases only to increase again on standing. This increase can be by as much as 200 times in a day. Figure 7.1 shows the plot of apparent viscosity against rate of shear, D, as determined by Price Jones[48] in a concentric cylinder viscometer. The curve shows a typical example of thixotropy (time dependent softening). Price Jones showed that this phenomenon was due to the 0·2–1·9% protein in the undetermined material. When this was removed the heather honey became Newtonian. When the protein extracted was added to Newtonian clover honey, this in turn became thixotropic.

The opposite effect is shown by Eucalyptus honey. It becomes stiffer on shearing and so it is dilatant or rheopectic. This behaviour may be so marked that when one places a finger into a jar of honey obtained from the flowers of *Eucalyptus ficifolia*, the apparent viscosity increases so much on pulling the finger away, that 2-m long

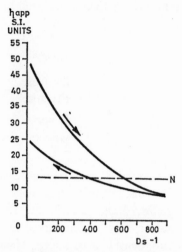

Figure 7.1 Plot of apparent viscosity against
shear rate for heather honey[48]

strands can be pulled off. These strands break rapidly indicating the time dependence of the phenomenon. It has been given the name 'Filamentation'. Figure 7.2 shows a plot of the apparent viscosity against shear rate D and the thickening is quite apparent. Unfortunately, no down curve has been prepared so it is not possible to see from this graph whether the material is time dependent or not. Again it was Price Jones who discovered in Eucalyptus honey some 7 per cent dextran, $(C_6H_{10}O_5)_n$ where $n \approx 8000$.

When he removed the dextran with acetone, in which it is soluble, the honey became Newtonian. When he added the extracted dextran to Newtonian honey this in turn showed dilatancy.

Eucalyptus honey also shows the Weissenberg effect to a marked extent (*see* Chapter 10) and must be presumed to show some elastic behaviour.

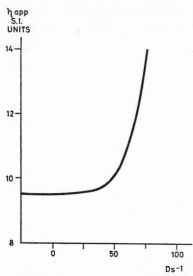

Figure 7.2 Plot of apparent viscosity against shear rate
rate for Eucalyptus honey (7·2% dextran)[48]

7.2 SOUPS

Rutgers has conducted an investigation on Dutch buttermilk soup,
containing 6–7 per cent pearl barley[50]. The soup is normally boiled
for 3 hours and this results in a thick creamy consistency. The soup
contains large, intact, swollen grains of the pearl barley 5–10 mm in
diameter. The investigation was conducted using the Haake Rotovisco[53] concentric cylinder viscometer. For exact work the gap between the cylinders must be narrow, about 1–2 mm. Such a narrow
gap was impossible because the oval barley grains were 4·5 mm in
diameter and 7–9 mm long. So Rutgers first studied the effect of the
gap. A gap of 7·6 mm and 29 mm was used and the readings were not
affected with thick and thin soups or starch and oatmeal soups.

Next the difference between smooth and rough surfaced rotating
cylinders was investigated. Although it is relatively simple to convert
the r.p.m. of the rotating cylinder into D, and the scale reading into
τ with Newtonian liquids, Rutgers did not do so because he felt that
the gap was too wide.

All his results were therefore plotted in r.p.m. versus scale reading and thus no confusion is caused. Table 7.1 shows some of the results obtained and a slight hysteresis is noticeable. This is expressed in the difference between the readings obtained on the upcurve and those on the downcurve.

TABLE 7.1

Thixotropy of buttermilk soup[50]

Scale reading

r.p.m.	Upcurve	Downcurve	Difference
3	6·8	5·3	−1·5
6	7·9	6·9	−1·0
18	9·5	8·8	−0·7
54	11·8	11·2	−0·6
162	14·9	14·3	−0·6
486	19·6	—	—

Thinning is also apparent from Figure 7.3 which shows the decrease in the scale reading (equivalent to apparent viscosity) with time at a constant revolution of 486 r.p.m.

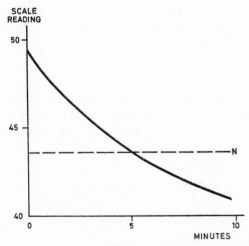

Figure 7.3 Thinning of buttermilk soup with time[50]

Figure 7.4 shows the τ–D and the η_{app}–D curves for cream of tomato soup as determined by the writer. The experimental procedure is given in Appendix A.5. It is apparent that the soup shows pronounced thixotropic characteristics.

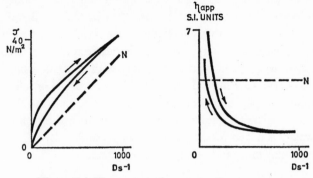

Figure 7.4 Flow curves for cream of tomato soup

7.3 HYDROCOLLOID SOLUTIONS

The most important practical aspect of carbohydrate gums is the way in which they change the rheological properties of food to which they have been added. There are very many of these gums and they are conveniently divided into:

(*a*) Exudate gums (gum arabic, gum tragacanth, gum karaya).
(*b*) Seaweed gums (agar, algin, carageenan).
(*c*) Seed gums (pectin, guar, corn hull and wheat gums, quince and locust bean gums).
(*d*) Starch and cellulose derivatives (amylose, amylopectin and substituted celluloses, such as methyl and ethyl cellulose).

All these are carbohydrates, macromolecular, and hydrophyllic. There is only one gum used in the food industry which is not a carbohydrate and is hydrophobic. This is chicle which is used in the manufacture of chewing gum.

All carbohydrate gums give highly viscous solutions or suspensions or gels with water. What then is the relationship between the structure of these polymers and their rheological properties in aqueous solutions?

Straight chain polymers

Since the molecules are hydrophyllic, they interact with immediately adjacent molecules of water and immobilize them. So the molecules assume a somewhat larger space due to their apparently enlarged diameter. If the polysaccharide exists in solution in an extended form with very little tendency to coil up, then as these stiff molecules gyrate in the solvent, they effectively sweep through a large spherical volume. Hence the apparent viscosity is relatively high in relation to molecular weight and increases rapidly with concentration. If the polysaccharides are neutral, there is little effect of pH or salts on the apparent viscosity unless complex formation results. As the linear polysaccharide molecules gyrate they may collide, shear off some of the water, and become adsorbed on one another. In this way the particles may grow and eventually precipitate.

Branched chain polymers

Branched neutral polysaccharides have a much lower apparent viscosity than the linear extended polymers of equal molecular weight, because they occupy much less volume than the linear polysaccharides (*see* Figure 7.5).

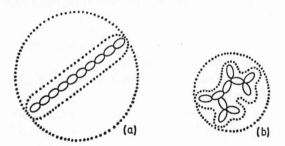

Figure 7.5 Diagrammatic representation[51] of: (*a*) straight chain; (*b*) branched chain polymers in solution. Both contain 10 units each

When these branched polymers collide they may become entangled and if the concentration is high enough a gel can result. Excessive association does not, however, occur since the molecules do not fit closely enough together and there is rarely spontaneous precipitation as with linear polymers. For the same reason they do not form strong films as straight chain polymers do but remoisten very quickly and are often used as adhesives.

Guar gum

Guar gum is of interest because it shows the properties both of a branched and of a straight chain polysaccharide polymer. Figure 7.6

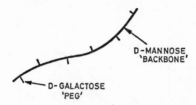

Figure 7.6 Diagrammatic representation of a guar gum molecule

shows the long backbone of D-mannose and small side chains of single molecules of D-galactose. Hence the molecule is very long but with very short side branches. It is highly viscous in solution as a linear polymer would be. It is also stable in solution and does not form precipitates. It redissolves easily, showing characteristics of the branched polymer. It also forms fairly strong and flexible films. Hence the structure of guar gum shows in an interesting way the differences between the branched and straight chain polymer.

Typical examples of the linear and branched type are pectin and gum arabic respectively. Their characteristics are shown in Table 7.2.

TABLE 7.2
Some characteristics of pectin and gum arabic

	Linear Molecule	*Branched Molecule*
Example	*Pectin*	*Gum arabic*
Apparent viscosity	High	Low
Precipitation	Easy	Not easy
Film strength	High	Low
Re-solution	Difficult	Easy
Flow of 10% solution	Softens	Newtonian

Although gum arabic has a very large molecular weight it appears to be coiled up to a roughly spherical shape and for that reason shows Newtonian properties in solutions up to 10 per cent. It has been noted (*see* Chapter 5) that milk is also largely Newtonian although it contains relatively large fat globules. As long as their interaction does not depend on flow rate the suspension will remain Newtonian.

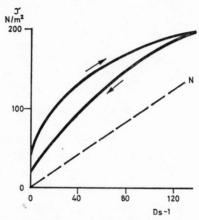

Figure 7.7 Thixotropic behaviour of high–calcium sodium calcium alginate solution[52]

Most aqueous solutions of carbohydrate gums show pseudoplastic characteristics but some show pronounced thixotropy. For instance, sodium calcium alginate at high calcium concentrations[52] will show this behaviour (*see* Figure 7.7).

8 Plasticity and its Measurement

8.1 PLASTIC FLOW AND THE BINGHAM MODEL

Typical plastic foods are, for example, mashed potato, whipped cream and nougat. These materials retain their shape under gravity. If, however, a force sufficiently greater than gravity acts on them, they flow almost like liquids. When the force is removed they retain their shape and cease to flow. In rheological terms a plastic material flows when a stress greater than a limiting value acts on it. This limiting stress is termed the yield value (τ_0). The ideal material which exhibits plasticity is the Bingham model (E. C. Bingham, 1878–1945, American scientist). For some practical purposes the criterion that a plastic material will not flow under gravity alone is used to distinguish a plastic substance from a liquid. It is in fact quite arbitrary. A spoonful of mashed potatoes on a plate will not flow under gravity, but a very large amount in a hopper will, simply because the weight of the material is great enough to cause flow in the lower regions of the tank.

It has been shown in Chapters 2 and 4 that the Newton and Hooke elements can be represented in three ways: as a model, as a shorthand symbol, and as a mathematical equation. The Bingham model is a complex of a Newton and a Hooke element together with a frictional slider. The slider is referred to as the St. Venant element, after the Count Barré de Saint Venant (1797–1886) and may be represented by a spring clip through which a piece of material is drawn. (The St. Venant element is also often shown as a block sliding over a surface.) The clip indicates that the material is rigid below the limiting friction (yield stress) and slides freely above it although still against the limiting frictional force τ_0. Figure 8.1 shows these three elements respectively in the form of a graphic model, the shorthand symbol and the equation.

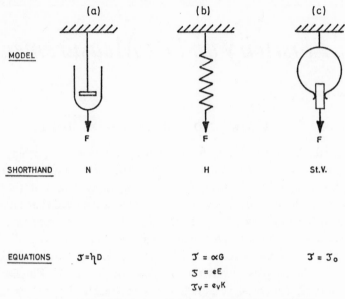

Figure 8.1 The Newton liquid (*a*), the Hooke solid (*b*), and the St. Venant slider (*c*)

The Bingham model

The Bingham model is a complex model consisting of a Hooke, a Newton, and a St. Venant element (*see* Figure 8.2). It is depicted as a spring clip and dash pot in parallel fitted to a spring in series. This is to indicate that under low stress, only the spring is operative, because no motion takes place in the other two elements, and the material will follow Hooke's law. If the stress exceeds τ_0, the spring clip will begin to slip and extension will then depend on the rate at which the Newtonian dash pot extends. If the stress falls again below the yield value, there is a slight elastic recovery due to the Hookean spring, but the deformation due to the dash pot will remain unchanged because there is no restoring force which acts on it. The spring clip holds it in place as long as the stress in either direction does not exceed τ_0.

(It should be noted that for all graphic representations which show elements in parallel such as the Bingham, Kelvin-Voigt and Burgers models, the horizontal bars must remain parallel during our imaginary experiment.) This behaviour is also expressed in the shorthand formula, where the Bingham model (Bi) equals the St. Venant element $(St.V)$ in parallel (indicated by a vertical line) with

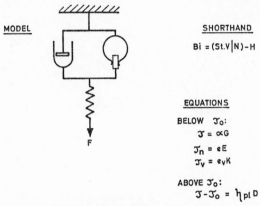

MODEL

SHORTHAND

$Bi = (St.V | N) - H$

EQUATIONS

BELOW $\mathfrak{I}_0$:

$\mathfrak{I} = \alpha G$

$\mathfrak{I}_n = \epsilon E$

$\mathfrak{I}_v = \epsilon_v K$

ABOVE $\mathfrak{I}_0$:

$\mathfrak{I} - \mathfrak{I}_0 = \eta_{pl} D$

Figure 8.2 The Bingham model

a Newton element (N). Both are in series (horizontal line) with a Hooke element (H). The equations for the Bingham model below the yield value are the same as those for the Hookean spring. Above the yield value, the equation is similar to that of a Newtonian dash pot except that the coefficient of viscosity is referred to as the plastic viscosity (η_{pl}) and the shear stress, τ, is effective only after subtraction of the limiting friction, τ_0, leaving the difference $(\tau - \tau_0)$ to act on the dash pot. The reciprocal of plastic viscosity is sometimes called *mobility*. It is equivalent to fluidity in Newtonian systems.

8.2 NON-IDEAL BEHAVIOUR

While the Bingham plastic is an ideal concept $(Bi$ in Figure 8.3), the $\tau - D$ diagram shows that η_{pl} is subject to the same complexities as the apparent viscosity (η_{app}) in addition to the complications of the yield value (τ_0). If the material is time independent (steady state) a power law may be applied as for pseudoplastic and dilatant materials.

$$\tau - \tau_0 = KD^n$$

It should be noted that if $\tau_0 = 0$ and $n = 1$, the material is a Newtonian liquid, and if $\tau_0 > 0$ and $n = 1$, the material is a Bingham plastic. The power law approach is subject to the same limitations as for non-Newtonian liquids (*see* Chapter 6). It should be noted that many real materials (suspensions, etc.) undergo small plastic deformations at low stresses but the movement stops quickly. When a value

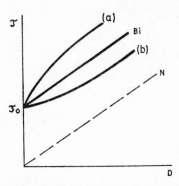

Figure 8.3 Shear stress τ plotted against shear rate D for ideal Newton (N) and Bingham (Bi) models. Curves (*a*) and (*b*) show thinning and thickening plastic systems respectively. τ_0 is the yield value

$\tau = \tau_0$ is exceeded, flow becomes continuous and in this region Bingham behaviour is shown. With such materials the 'yield point' is not sharply defined.

8.3 STRUCTURE OF PLASTIC MATERIALS

Just as there are two types of elasticity, namely, ordinary and high or 'rubber' elasticity (*see* Chapter 2) so there are two types of plasticity. By analogy we can call the first ordinary plasticity or 'metal' plasticity (because it occurs mainly in metals) and the second high plasticity.

In a metal, the electrons of the outer atomic shell are shared amongst the atoms to give rise to what has sometimes been called an 'electron gas'. When deformation of the metal exceeds the elastic limit, a permanent displacement of the atoms against the interatomic forces occurs and the new equilibrium is nearly as good as the old one although some distortion and faults may arise in the structure. The yield value is of the order of 7–40 × 10^7 N/m². In food science we are not concerned with this type of plasticity in the actual foodstuffs.

High plasticity is quite different and the yield value is usually of the order of only 10–100 N/m². Just as with high elasticity it has been suggested that a material must have three characteristics for it to occur (*see* Chapter 2), so with high plasticity there appear to be three underlying structural factors.

If we observe a sample of margarine under the microscope, we discern a mass of solid crystals in a continuous phase of oil. This combination is typical for a plastic material of this type. Three conditions must apply:

(*a*) There must be a two-phase system consisting of a continuous liquid phase and a 'solid-like' dispersed phase. This 'solid-like' phase need not necessarily be truly solid as long as it acts like a solid. For instance a bubble of gas or liquid may act as a solid, if surface tension makes it behave so. In whipped cream or similar foams the 'solid-like' phase is a gas with its surface film. In mayonnaise or similar emulsion, the 'solid-like' phase is a liquid. A true yield value cannot, by definition, be found in a single phase liquid (*see* Table 8.3).

(*b*) There must be a fine dispersion of solid in liquid. There must be no seepage or settling and the whole mass must be held together by internal cohesion.

(*c*) There must be a correct proportion of 'solid-like' to liquid phase. If there is too much solid, there is a rigid interlocking and the material becomes brittle. If there is too much liquid, the material will flow under gravity and does not exhibit a yield value: the solid does not jam or interlock. Table 8.1 summarizes plastic and elastic behaviour.

Although there must be a correct proportion of 'solid-like' to liquid phase, the proportion below which flow occurs varies widely with different materials. Mashed potatoes for instance contain 90 per cent water and chocolate melt only 35 per cent of liquid fat, but both materials are plastic.

TABLE 8.1

Characterization of plastic and elastic phenomena

	Ordinary elasticity	*High (rubber like) elasticity*
Young's modulus	10^{10}–10^{11} N/m^2	10^5 N/m^2
Mechanism	Interatomic displacement	Displacement of chain molecules
Conditions		1. Long chain molecules 2. Weak intermolecular forces 3. Firm cross links at few points

	Ordinary (metal) plasticity	*High plasticity*
Yield value	7–40×10^7 N/m^2	10–100 N/m^2
Mechanism	Interatomic displacement	Phase displacement
Conditions		1. 2-phase system 2. 'Solid-like' dispersion in liquid 3. Correct proportion of solid in liquid

The yield value τ_o, is not related to the plastic viscosity η_{pl}, and η_{pl} has no direct relation to the viscosity of the liquid phase. This is clearly shown in Table 8.2.

TABLE 8.2

Plastic viscosity (Pl) and yield value ($\tau_0(N/m^2)$))
at 30°C for various foods[54,55]

	η_{pl} (*Pl*)	τ_0 (*N/m²*)
Catsup	0·08	14·0
Mustard	0·25	38·0
Mayonnaise	0·63	85·0
Oleo margarine	0·72	51·0
Protein foam	1·0	40·0
Honey	11·0	0

TABLE 8.3

Various examples of plastic foods with particular
reference to their solid-like and liquid phases

Example	*Solid-like phase*	*Liquid phase*
Marzipan	Sugar, almonds	Sugar solution
Margarine	Fat crystals	Oil
Mayonnaise	Oil and surface film	Aqueous solution
Whipped cream	Air and surface film	Protein solution

8.4 PLASTICITY MEASUREMENT

Measurements on plastic materials can be taken either in the elastic region, below the yield value, or in the plastic region above it. There are not many studies of the former type, because measurements in this region do not appear to be technologically important. In addition few real materials behave in an ideally elastic way in this region. The techniques used can sometimes be the same as those for solids but obviously the apparatus must be used at very low stress, below τ_0, so the concentric cylinder viscometer can be used conveniently to study the elastic behaviour. In this way the elastic moduli G and E can be obtained. If there are no gases present, K will be large, as for liquids, and, if needed, can be measured in the same way.

In the plastic region above τ_0, two types of fundamental measurement have been used, the capillary viscometer and the concentric cylinder viscometer.

(a) Capillary viscometer

When considering Newtonian viscosity in Chapter 4, we saw that the shear stress was highest near the tube wall and zero at its centre. Therefore somewhere near the centre, shear stress must be equal to τ_0. Where the shear stress is below τ_0, there is no flow and therefore in all plastic materials passing along a tube there must be near the middle of the tube a solid core, the diameter of which depends on the shear stress (the greater the shear stress, the smaller the diameter of the core). Hence the flow curve ($\tau - D$) for a plastic material is never a straight line, not even for the ideal Bingham model because the amount of material sheared depends on τ. Nevertheless, there is an approach towards a straight line as τ is increased.

Since τ_0 is finite and τ at the tube axis tends to zero, there is always a core with plastic materials. This core can be seen if we observe plastic materials flowing through a capillary under the microscope (*see* Figure 8.4). The technique given in Chapter 6 for obtaining

(a)   (b)

CORE

Figure 8.4 Stream-line flow of: (*a*) a Newtonian liquid;
(*b*) flow of a plastic material

P^* and the D value corresponding to it can be used for plastic materials and is especially useful where they do not show ideal Bingham behaviour (*see also* section 9.3).

The Bingham equation states that:

$$\tau - \tau_0 = \eta_{pl}D$$

Reiner and Buckingham have elaborated this equation to deal with the flow of plastic materials through a capillary. The equation applies only to the ideal Bingham material, but it is still worth trying in practice if it is suspected that the material is not time dependent and resembles the Bingham plastic closely. The Buckingham–Reiner equation states that:

$$\eta_{pl} = \frac{\pi r^4 p}{8lQ}\left[1 - \frac{4}{3}\left(\frac{2l\tau_0}{rp}\right) + \frac{1}{3}\left(\frac{2l\tau_0}{rp}\right)^4\right]$$

η_{pl} is the plastic viscosity (N_s/m^2), τ_o is the yield value (N/m^2), r is the capillary radius (m), l is the capillary length (m), p is the pressure difference (N/m^2), and Q is the flow (m^3/s).

For continuous measurement of the yield value the Eolkin Plastometer is available which works also on the capillary principle[56].

(b) Concentric cylinder viscometer

In a concentric cylinder viscometer with a rotating inner cylinder the shear stress is greatest near the wall of the inner cylinder and tends to zero near the wall of the outer one. As in the capillary viscometer there is a region where τ lies below τ_o. This can be demonstrated by drawing a straight line on the surface of a plastic material in a concentric cylinder viscometer using a wire dipped in carbon powder.

After a few rotations with a plastic material in the gap, the line disappears in the region of shear. The region where τ is below τ_o lies near the outer wall and is equivalent to the core in the capillary tube (*see* Figure 8.5).

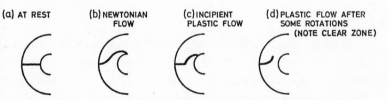

Figure 8.5 Flow in a concentric cylinder viscometer

Just as Buckingham and Reiner have developed an equation suitable for capillary tubes and based on the Bingham equation, so Reiner and Riwlin have produced an equation for use with plastic materials in concentric cylinder viscometers[3].

8.5 EMPIRICAL MEASUREMENT

Empirical measurements on plastic materials have been borrowed either from solids or from liquids testing. There are very many methods and five of them will be briefly considered here.

(a) Penetrometers

In the first type of penetrometer a needle is dropped through a guide tube onto the sample and the penetration is measured (*see* Figure 8.6(*a*)). The range of this instrument is limited. If the sample is too stiff there is too little penetration to measure accurately; if it is too soft, the needle continues to descend.

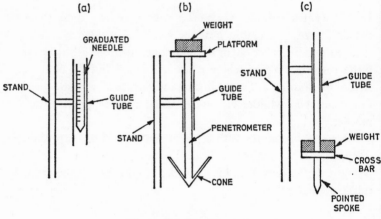

Figure 8.6 Penetrometers

Alternatively a cone is allowed to descend into the plastic material, either by constant weight or constant rate of descent (motor). Sometimes a spherical indentor is used. In this type of measurement the cone surface, the cone angle, the kinetic energy and the penetration time all affect the reading (*see* Figure 8.6(*b*)). A very simple type of penetrometer is shown in Figure 8.6(*c*). Here a bicycle spoke has been sharpened at the end and a small cross bar has been welded to it. A weight is dropped repeatedly on to the cross bar, the instrument acting like a miniature pile driver. The first drop is neglected because it may contain an error due to surface irregularities of the plastic material. The penetration is plotted against the number of drops.

(b) Compression of cylinders

This is not a good test for plastic materials because a slight increase in force will result in a very great increase of deformation. Furthermore the surface on which the pressure acts increases as the cylinder increases in diameter on compression.

(c) *Extrusion*

There are mechanical or pneumatic devices which allow the material to be extruded under positive or negative pressure. A common instrument is the FIRA/NIRD extruder[57] where a piston moves at uniform speed to extrude the material through an orifice in the bottom of a cup. The cup is suspended and forced against a spring actuating a pen which records the thrust in kg on a moving chart. Extrusion using tubes can be adapted to allow the use of the equations for flow through a tube.

(d) *Sectility*

In this method the sample is cut by a wire, much as the grocer cuts a cheese. The taut wire may be moved down by a motor and the force exerted by the sample upwards is measured. Alternatively, the sample may be moved upwards and the force on the wire measured. In either case the movement should be slow and of the order of 0·5 mm/s.

(e) *Sagging beam*

In this method the bar-shaped sample is placed on two knife edges and the sag determined in the middle. This method has been used for fats, where 6 hours were allowed at a controlled temperature of 20°C.

8.6 STICKINESS

Stickiness is a surface property of viscous or plastic materials and is conveniently dealt with in this chapter. It depends on the cohesion of the sample itself, and the adhesion of the sample to another body. If cohesion is lower than adhesion, the sample ruptures. If the reverse is the case, the sample separates from the other body. Claasens[58] has used the term 'hesion' to include both adhesion and cohesion.

If two discs are held together by an adhesive film and moved apart from l_1 to l_2, there is a flow of film at right angles to the separating

4

force. This flow depends on the viscosity of the film. The Stefan equation[59] states that:

$$\tau_n t = \frac{3\eta r^4}{4} \left(\frac{1}{l_1^2} - \frac{1}{l_2^2} \right)$$

where τ_n is the normal stress, t is the time, η is the coefficient of viscosity, r is the radius of the discs, l_1 is the original and l_2 the final distance separating the discs. Figure 8.7 shows a diagram of this arrangement.

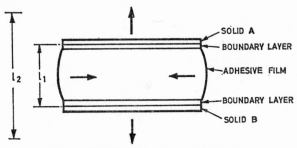

Figure 8.7 The flow of an adhesive film

The Stefan equation has considerable limitations. It neglects surface roughness, solid-liquid forces, and non-Newtonian behaviour, but nevertheless it places stickiness on a more intelligible basis.

Many foods stick readily to our fingers which are relatively dry, (dough, wheat gluten, chewing gum, syrup) or to our teeth which have a wet surface. The latter phenomenon is noticeable by the difficulty experienced in separating the teeth once we have bitten onto food such as toffee, masticated bread or meat. All these materials act as liquids or plastics of relatively high viscosity showing hesion between themselves and the teeth or fingers.

Hesion is determined in the laboratory by pulling a disc or plunger from the surface of the material. A simple apparatus which is easily built in the laboratory is shown in Figure 8.8.

A simple balance bears a pan and a beaker on one side, and a disc fixed at right angles to the supporting wire, on the other. Additional discs may be added to the plunger to increase the weight on the sample. On running water or mercury into the beaker, the load on the sample decreases and eventually the plunger is torn away. It is important that the pull should take place at right angles to the surface, that the plunger should be clean, and that statistics are applied to the results because these are very variable. For boiled sweets, Heiss[60] has used a parallel plate apparatus, where a flat boiled sweet was compressed for 10 s at 325 g. The lower plate was then withdrawn at a constant

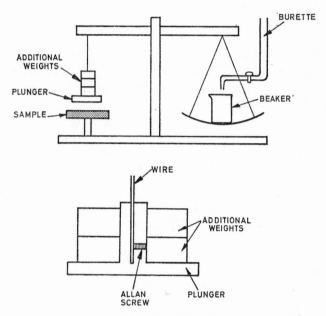

Figure 8.8 A hesion balance

speed. The upper plate was attached to a spring and when it was pulled away, the distance along which the lower plate had travelled was read. This gave a measure of stickiness.

A similar instrument was used by Autard for cake batters[61].

There is relatively little information on the factors affecting stickiness. Thomasos and Wood[62] have shown that the stickiness of butter is affected mainly by the discontinuous phase, by the size of the butter fat crystals, and any gas occluded in the material. Muller[63] showed that stickiness between dough and the paddle blades of the Farinograph affected the consistency curve to a considerable extent.

9 Plastic Foods: Examples

It was pointed out in Chapter 1 that all materials exhibit all rheological properties (if the conditions are appropriately chosen) but some predominate in practice. This is particularly true for plastic materials which on the whole are very complex indeed. Foams, for instance, are often regarded as visco-elastic if their yield value is very low. Fats when hardened are often quite solid but when aerated, are visco-elastic. Nevertheless, foams, chocolate melt, and fats have been chosen in this section to illustrate plasticity. This approach is justified because yield value and plastic viscosity are their most significant technological criteria. It is, however, important to remember that these are not the only rheological features of these substances.

9.1 FOAMS

When egg white is beaten in air, a foam is produced. The beating action causes ridges, which at first flatten out rapidly due to the viscous properties of the egg white. If beating is continued, the ridges remain at first as soft rounded 'peaks' then as hard sharp 'peaks'. At this point, the foam will no longer flow under gravity. A yield value has been reached and the foam is now plastic (*see* Figure 9.1).

Plastic egg foam shows the three conditions for plasticity mentioned in Chapter 8. First, there is a two-phase system: the continuous phase is protein solution and the solid-like phase is the air bubbles with their surface film. Secondly there is a fine and critical dispersion of air bubbles in the liquid. Thirdly, there is a correct proportion of 'solid-like' to liquid at 'peaking'.

It seems that the rheological properties of a foam depend primarily on the viscosity of the continuous phase, and on the size of the bubbles.

The latter plays the predominant role in determining the yield value. There is little quantitative work on foams but it is known that τ_0 depends on the type of foam, its age, and previous history.

Two methods have been used to determine quantitatively the yield value of protein foams[65]. The first is the 'wafer method'. This is used for very stiff foams (protein foams). The foam is placed

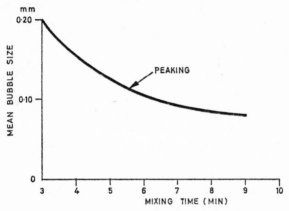

Figure 9.1 The relationship between bubble size and mixing time for egg white[64]

between two rectangular plates of zinc gauze and a pull is exerted on the top gauze until it yields. τ_0 is calculated from the force on the gauze and its unit area. The second is the 'cylinder method'. Here an instrument similar to a concentric cylinder viscometer is used. The foam is placed into the outer vessel and the inner cylinder suspended from a steel wire. The outer cylinder is slowly rotated until the inner one ceases to follow. Knowing the elastic constants of the wire, the value of τ_0 can be calculated from a pointer on the inner cylinder which indicates the maximum deflection.

9.2 MELTED CHOCOLATE

Chocolate consists of finely ground sugar and cocoa solids dispersed in cocoa butter. Cocoa butter melts just below the temperature of the mouth and there releases the flavouring materials of the chocolate. Milk chocolate contains additionally butter fat and non-fat milk solids.

When an embossed bar of chocolate is heated to about 80°C on a tray, its shape remains unchanged although the chocolate has softened. If we give the tray a sharp tap, the chocolate will suddenly flow because the shock has exceeded the yield value of the chocolate. The plastic properties of melted chocolate are of considerable importance industrially, because chocolate is often deposited into moulds or used to coat centres. If the chocolate is too thick, air bubbles will not easily be removed by the shaking equipment, and the chocolate coating becomes too thick. If the chocolate is too thin, the coating will be too thin, and the chocolates will develop base flanges and poor decorative marking. The detailed rheology of chocolate is not clear, but shows a complex softening or hardening under specific conditions.

The first experiments on chocolate were conducted with a capillary viscometer which was not particularly suitable. About 1950 a concentric cylinder viscometer was used and it was found that the $\tau - D$ relationship was roughly a straight line, but the curve did not pass through the origin. It resembled a Bingham plastic:

$$\tau - \tau_0 = \eta_{pl} D$$

Better results were obtained by using the Casson equation which had been proposed initially to deal with the rheology of flocculated particles. The equation states that:

$$\sqrt{\tau} - \sqrt{\tau_0} = \sqrt{\eta_{pl}}\sqrt{D}$$

The line is reasonably straight and its slope gives η_{pl}. The intercept on the stress axis determines the yield value. Recently, experiments showed that the apparent viscosity of molten chocolate depends not only on the rate of shear but also on its duration and the chocolate does not recover its original viscosity after stirring[66].

It was felt that the Casson equation was not applicable over the whole range of behaviour investigated. This was particularly true for milk chocolates and plain chocolate containing polyglycerol polyricinoleate.

The equation proposed adds an additional term to the Bingham equation and results in three material constants:

$$\tau - \tau_0 = \eta_{pl} D + \text{B} \sin h^{-1} D$$

where τ is the shear stress, τ_0 is the yield value, D is the shear rate, η_{pl} is the plastic viscosity and B is a constant (sin h is the hyperbolic sine).

Some workers have favoured the Williamson equation for chocolate, which will be considered in the following section on fats.

η_{pl} is affected to a considerable extent by processing, i.e. degree of roasting of beans, conching or ageing. 0·3–0·4 per cent of lecithin has a considerable effect on the flow properties of chocolate. While plastic viscosity is continually decreased with increasing levels of lecithin, τ_0 passes through a minimum at about 0·5 per cent lecithin and is then increased again. The effect of lecithin on the plastic viscosity depends therefore on D (*see* Figure 9.2).

The mechanism of the effect of lecithin is not known but it is important that within certain limits both τ_0 and η_{pl} can be altered to suit manufacturing conditions.

It has been stated in Chapter 6 when introducing the concept of apparent viscosity, that the $\tau - D$ graph does not give a straight line. It was shown that for such systems, one point measurements are no longer adequate because they do not characterize the flow curve. Here is an exception: Figure 9.3 shows curves for several chocolates of the same type. Since the curves are similar it is apparent that a single point measurement will in fact characterize each sample very well indeed. This is an illustration of the maxim that one may break almost any rule as long as one acts from knowledge.

9.3 FATS

Below the yield value, fats show elastic properties[68], but it appears that these are not of great technological importance. Above the yield value, both the yield value magnitude and the plastic viscosity are of importance. In bakery technology, fats are frequently employed and it is of importance that they be plastic. If a hard fat is used, this is found as lumps in the final baked product. If an oil is used, cakes will not rise adequately. It is largely due to crystal structure and arrangement that air is retained on mixing a plastic fat and a satisfactory cake volume results. The importance of the yield value in the blending of margarine is well known: 'It spreads straight from the 'fridge'. This is one of the few advertising slogans where the rheological properties of the material are used to sell the product. Table 9.1 shows the psychological assessment of fat in relation to its yield value.

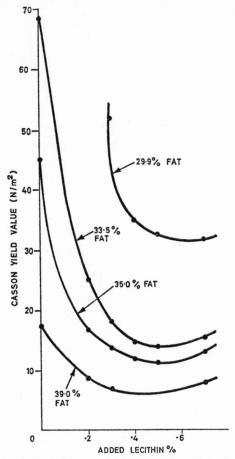

Figure 9.2 The effect of lecithin on the yield value of plain chocolate[67]

The rheological properties of a fat depend on the temperature of storage, any pasteurization, cooling rate after pasteurization, work, and in the case of butter, churning and the temperature of the wash water.

The work of Søltoft[41] in relation to oils has already been mentioned (*see* Chapter 5). He also studied the rheology of fats using a capillary

Table 9.1

Psychological assessment of fat according to yield value[69]

τ_0 ($\times 10^3$ N/m²)	Assessment
< 5	v. soft, just pourable
5 – 10	v. soft, not spreadable
10 – 20	soft but just spreadable
20 – 80	plastic and spreadable
80 – 100	hard but spreadable
100 – 150	v. hard, just spreadable
> 150	too hard to spread

Note : To obtain results in dyne/cm² multiply by 10

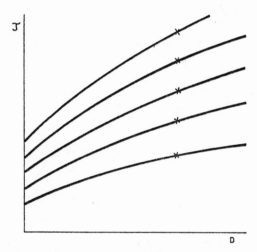

Figure 9.3 Family of curves for different samples of chocolate

tube viscometer with compressed air to force the material through the capillaries. He used peanut oil, and peanut oil hardened to melting points of 30°, 40° and 50°C. His experiments conducted at 50°C, when the peanut oil was liquid, are reported in Chapter 5, Table 5.7.

At 16°C all these materials were solid and with one exception showed plastic flow and a yield value. Søltoft's plot is a little unusual.

He had to use various tube dimensions because of the various consistencies of the fats. To make his results comparable, he plotted the shear stress at the tube wall $pr/2l$ against the rate of deformation $4Q/\pi r^3$. Figure 9.4 shows his results for various mixtures of peanut

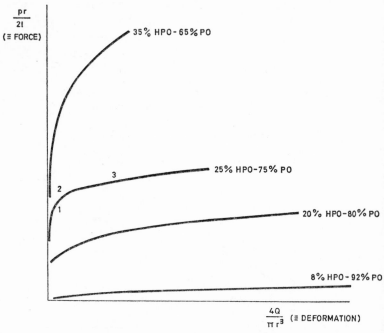

Figure 9.4 Flow curves of fats[41] at 16°C

oil and hardened peanut oil. It is apparent that only one of the curves shows largely Newtonian behaviour, while all the others show a yield value and a plastic viscosity.

It has been pointed out in Section 8.2 that with some real materials the 'yield point' is not sharply defined. 25 per cent hardened peanut oil and 75 per cent peanut oil mixture is typical of such a plastic solid.

There are three parts to the curve: (1) a region of slow flow; (2) a curved transition range; and (3) Newtonian liquid flow. Each part of this curve can be extrapolated graphically to meet the force axis and each one will result in a slightly different yield value.

It should be noted that if $P^* = pr/2l$ is plotted against $Q^* = 4Q/\pi r^3$ for different lengths and diameters of tube, a unique plot would indicate that there is no wall slip and probably no thixotropy (*see* Sections 6.4 and 8.4).

Figure 9.5 shows the rheological behaviour of butter and a well blended margarine at 20°C. The yield value in the margarine is

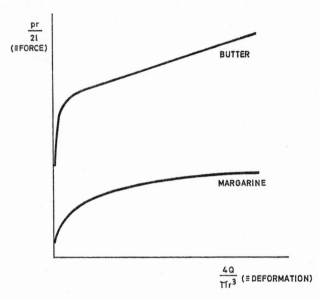

Figure 9.5 Flow curve of butter and margarine[41] at 20°C

much less pronounced and rheological properties are not affected so much by temperature as those of butter.

Søltoft found that the equation which best fitted his results was that of Williamson. This equation had been developed on the assumption that in dispersed systems, part of the shearing force is used to break down the structure, while the remainder is used to produce flow at higher shear rates. The equation is an empirical power equation and has no theoretical basis.

There are many forms of the Williamson equation and the one that Søltoft used was of the form:

$$\frac{pr}{2l} = K_1\left(\frac{4Q}{\pi r^4}\right) + \frac{K_2\left(\dfrac{4Q}{\pi r^3}\right)}{K_3 + \left(\dfrac{4Q}{\pi r^3}\right)}$$

In this equation p is the pressure difference, r is the capillary radius, and l is capillary length. Q is the volume delivered and K_1, K_2, and K_3 are constants. So the behaviour of fat was very dissimilar to that of the ideal Bingham plastic, and as expected, the Buckingham–Reiner equation did not show a good fit to Søltoft's experimental results.

10 Visco-elasticity and its Measurement

10.1 VISCO-ELASTICITY

If wheat flour and water are mixed in a suitable proportion, a dough is formed. It is then easy to obtain a strip of dough, either by extrusion or by the use of scissors. If that strip of dough is placed on a pool of mercury (*see* Chapter 11) it will float. If the ends of the dough are now pulled gently, the dough will stretch and then it will appear to flow like a viscous liquid. If the ends are released, the dough will contract like a piece of soft rubber, although elastic recovery is only partial: the dough will not quite resume its original length.

This experiment shows that dough exhibits simultaneously both the viscous properties of a liquid and the elastic properties of a solid. Dough is visco-elastic.

In the plastic materials discussed in the last two chapters, we found elastic properties dominating the behaviour at small stresses, and viscous properties at large stresses. The two regions of behaviour were separated by the yield value τ_0. Although the elastic strain is still present above the yield stress, it is negligible compared with the plastic flow. In visco-elastic materials, the viscous and elastic components express themselves at the same time and under the same conditions. Elasticity is particularly readily detectable at high stresses and in contrast with the behaviour of plastic materials.

10.2 VISCO-ELASTIC MODELS

The simplest visco-elastic model can be represented by one spring and one dash pot, but there are two ways in which these two components can be arranged: in series and in parallel.

The former which is essentially a liquid, is referred to as the Maxwell model, the latter which is a solid as the Kelvin-Voigt model. These two models may in turn be joined together in series to give a model which closely resembles very many food substances. This is known as the Burgers model.

(a) The Maxwell model

The Maxwell model is called after James Clerk Maxwell, a Scottish physicist (1831–1879). It consists of a spring and a dash pot joined

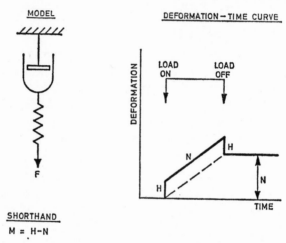

Figure 10.1 The Maxwell model

in series and represents the ideal visco-elastic liquid because the slightest force will cause it to flow. Figure 10.1 shows the spring and dash pot representation, the shorthand formula, and the deformation–time curve. The latter shows the increase in length of the model on stretching and the subsequent partial contraction when the load is removed.

When a load or force, F, is applied, H extends immediately. Then N elongates slowly. On the removal of the force, H contracts but N remains unchanged. This is expressed in the deformation–time plot which shows the deformation of H and N on application and removal of the force F.

There are four points in connection with the Maxwell model which are worth remembering:

(*a*) N has a coefficient of viscosity, and H has three moduli G, E and K. Usually the model can be regarded as incompressible so that no account need be taken of K and, if isotropic, then $G = \frac{1}{3}E$.

(*b*) At constant stress, H will instantaneously extend to its appropriate deformation for that particular stress. After that, the deformation depends on η and in this respect the Maxwell liquid behaves as a Newtonian liquid.

(*c*) If the model is extended and *held* (i.e. constant strain), the internal stress relaxes, i.e. H pulls N out until the stress has disappeared. This process is called relaxation. The ratio η/G is called the 'relaxation time' T_M.

(*d*) If the constraint is released before relaxation is complete, recoil occurs. These properties cause flow instability and result in a low critical Reynolds number (*see* Chapter 4). Recoil and relaxation are typical of visco-elastic fluids and do not occur in viscous fluids.

(*b*) *The Kelvin-Voigt model*

This model is usually called after Kelvin (William Thomson, later Lord Kelvin of Largs, British physicist, 1824–1907) in Britain, and after Waldemar Voigt (German physicist, 1850–1919) in Germany. In the interest of Anglo–German relations, particularly dear to the heart of the author, it will be called by both their names in this book.

In the Kelvin–Voigt model the spring and dash pot are joined in parallel. The model represents a solid. The limiting extension after sufficient time lapse depends on the magnitude of force only and is completely recoverable when the load is removed (*see* Figure 10.2).

When a load or force, F, is applied, both H and N extend. When the force is removed, H pulls N slowly together again. This results in the phenomenon referred to as 'delayed elasticity' or 'elastic after-effect'. This is defined as the delayed recovery from elastic strain. In this model complete recovery theoretically takes an infinite length of time, but the time for effective recovery depends on the relative size of the constants of the viscous and elastic elements.

Unlike the Maxwell model, the Kelvin–Voigt model has its two elements in parallel. In the former both elements take the same stress like the links of a chain. The total elongation is therefore the sum of the elongation of each element. In the Kelvin–Voigt model the elements are in parallel and there is the same strain for both elements.

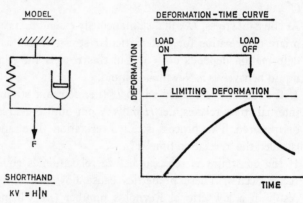

Figure 10.2 The Kelvin–Voigt model

The stress is the sum of the stresses of each element. Therefore it is clear that although both the Maxwell and the Kelvin–Voigt models have only one H and one N each, the equations for the two systems must be different. Points to remember for the Kelvin–Voigt model are:

(a) The total stress is the addition of the stress on the N and H elements.

(b) The strains for N and H are equal.

(c) The limiting behaviour of the Kelvin–Voigt model after sufficient time is that of the Hooke model.

(d) η/G is referred to as the retardation time T_{KV}.

(c) *The Burgers model*

In the food industry, materials resembling the Maxwell and Kelvin–Voigt model are hard to find. If however, these two models are joined in series, the resultant model, called after J. M. Burgers, a Dutch rheologist, is of considerable importance. This model is shown in Figure 10.3.

When the load or force, F, is applied to this model, there is an immediate deformation of H_2. This is referred to as instantaneous elasticity. The Kelvin–Voigt model is responsible for the delayed elasticity and the elastic after effect (H_1, N_1). After sufficient time, extension is solely due to the viscous part of the Maxwell model (N_2). On unloading, H_2 causes the instantaneous recovery, and the delayed recovery

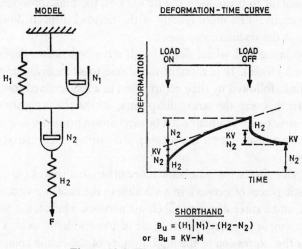

Figure 10.3 The Burgers model

is due to the Kelvin–Voigt part. Rapid and delayed recovery together are referred to as total elasticity, the remainder (N_2) is the viscous deformation.

The various equations for the Maxwell, Kelvin–Voigt, and Burgers models are beyond the scope of this book and the reader is advised to study Reiner's *Deformation, Strain and Flow*[3]. If he still thirsts for knowledge, he should consult V. Wazer *et al.* (*see* Chapter 6: Visco-elasticity and its measurement[31]). Then he could read Chapter 11 by Turner Alfrey and Gurnee entitled 'Dynamics of Visco-elastic Behaviour', in Eirich's *Rheology*, Volume 1[70]. If all these still leave him dissatisfied, he could well be in a position to write his own book.

10.3 STRUCTURE OF VISCO-ELASTIC MATERIALS

Visco-elastic materials range from the viscous liquid with elastic properties (Maxwell type) to the elastic solid with viscous properties

(Kelvin–Voigt type). Under normal conditions, elasticity due to the stretching of interatomic bonds is not noticeable and it is sufficiently true to say that all elasticity in visco-elastic foods is due to the elastic deformation of large molecules (*see* Section 2.5). In visco-elastic foods there is often only partial recovery because the elastic three dimensional network breaks down under stress. The more complete and resistant to breakdown the network is, the more pronounced is the elasticity. The more readily is the network broken down, the greater is the viscous component.

In gelatine and wheat flour dough stress relaxation occurs (*see* Maxwell model). It is thought to be due to successive rupture of cross links followed by their reformation in a less strained configuration. In this way the stress disappears, whilst the integrity of the whole structure is maintained. The mechanism primarily responsible for relaxation in dough is probably a disulphide–sulphydryl interchange.

Extremes in visco-elastic behaviour can be visualized as a dispersion of elastic pieces of network in a solution in the case of a visco-elastic liquid, and a three dimensional elastic network which tends to break down more or less readily in the case of a visco-elastic solid. Viscous flow is the expression of the discontinuity of the elastic component.

10.4 VISCO-ELASTIC MEASUREMENTS

There is a large range of visco-elastic materials and hence a large range of testing techniques. In contrast to plastic materials where the viscous and elastic components are effectively separated by the yield value, in visco-elastic materials both components usually express themselves together. It is, however, possible to separate the two components by carefully designed techniques and four of these are known.

(a) Vibrational measurement

Imagine a Maxwell liquid flowing along a capillary tube viscometer. Under sustained flow conditions the individual elements of the fluid have a constant deformation due to the elastic component and flow is entirely due to the viscous component. The elastic element is not apparent, but becomes so, if the stress or the flow rate suddenly change. (In turbulence there is a continuous unsteady state and elastic properties contribute markedly to turbulence—*see* Chapter 4.) This technique of varying the stress or flow rate is made use of in vibrational measurements where an alternating cycle is applied repeatedly to either stress or strain. The liquid is placed for instance between the cone and plate of a viscometer and instead of being subjected to rotation, the cone is oscillated back and forth. If the amplitude of the cone, the input, is plotted against time, a regular wave (sine wave) is obtained. If the oscillation of the plate, the output, is plotted similarly, its response is both delayed and its movement not as vigorous as that of the cone. The response is said to be both 'out of phase' and 'damped' in relation to that of the cone. From these two effects, viscous and elastic properties can be calculated. Since both are determined under dynamic conditions, the constants are referred to as the *dynamic* modulus and the *dynamic* coefficient of viscosity (*see* Figure 10.4), and they correspond to a particular frequency.

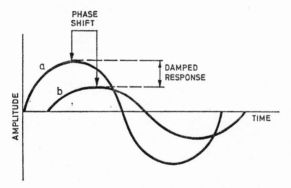

Figure 10.4 Oscillation of a visco-elastic liquid
(*a*) input, (*b*) output signal[35]

(*b*) *Loading-unloading experiments*
Figure 10.5 shows the mercury bath extensometer which has been used for visco-elastic measurements on wheat flour dough[72,73] (*see also* Appendix A.8). A dough cylinder floating on mercury is stretched by the application of a weight and its increase in length is measured by a ruler. After a certain time, the load thread is burnt with a match

and the unloaded dough contracts. Figure 10.6 shows the deformation–time curve.

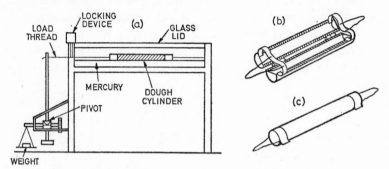

Figure 10.5 (*a*) The mercury bath extensometer, (*b*) the hinged knife, and (*c*) the dough cylinder[86]

This curve during the loading period is referred to as a creep curve. Creep is defined as the increase of deformation with time, and creep recovery as the decrease of deformation with time after unloading.

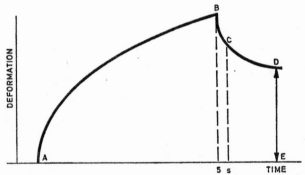

Figure 10.6 Deformation–Time curve for wheat flour dough

It will be noted that this curve is closely similar to the deformation–time curve of the Burgers model depicted in Figure 10.3 but it does not show immediate elasticity at *A*.

Furthermore *BC* in Figure 10.6 is referred to as rapid elasticity. It is the elastic recovery taking place within 5 s. Hence it is not quite the same as the so-called 'instantaneous' elasticity which proceeds theoretically with the speed of sound and is found in the ideal Burgers model.

Very many graphs of this type (*see* Figure 10.6) have been described in the literature and one of their most important characteristics is the description of the vertical axis (ordinate).

The most common designations on this axis are as follows:

(a) Deformation, (extension or compression) in terms of units such as cm, inch or angle. This is an empirical measure since the original length is not given.

(b) Strain, (relative deformation). There are two types of strain. If the original length, L, and each increment measured, l, are given, then the length at any time of the experiment is $L + l = L'$. l/L is referred to as the Cauchy measure of strain. l/L' is referred to as the Swainger measure. Allowance must be made for the decrease in cross sectional area when calculating the stress. The Swainger strain is often used if the extension is relatively large and as a result the decrease in cross sectional area considerable.

(c) Compliance. Creep compliance is the strain divided by the shear stress at any time during the experiment and it is perhaps the most meaningful parameter. If the deformation–time curves taken at different stresses are linearly related (doubling of stress should give a doubling of strain at the same time), the visco-elastic behaviour is said to be linear. 'Linear' here refers to a relationship being expressed by a linear differential equation. This is an equation in which the stress and strain components and their derivatives appear only as the first power. For instance Hookean springs, Newtonian dash pots and any combination of these including the Kelvin–Voigt and Maxwell models are linear. Any combination containing the St. Venant element is not. In practice, few foodstuffs give linear relationships. Non-linear visco-elasticity will be dealt with in the section on dough (*see* Chapter 11).

(c) *Stress relaxation method*

It has been shown earlier in this chapter that if a Maxwell model is extended and held, the internal stress relaxes. In a stress relaxation experiment the specimen is suddenly deformed and the disappearance or 'decay' of stress is recorded (*see* Figure 10.7).

There are several instruments which can be used for stress relaxation experiments such as the co-axial cylinder viscometer[74] or the various extensometers[75,76]. The most important value obtained is

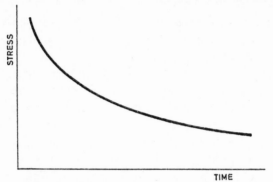

Figure 10.7 Stress decay at constant strain

the relaxation time T_M, i.e. the ratio η/G. If the log of the stress plotted against time results in a straight line, then the behaviour of the material is Maxwellian and T_M is given by the slope of the line. If the plot is not a straight line, and this is usually the case, rheological behaviour must be represented by several 'Maxwell' models in parallel but only if the system is linear (*see* section 10.4(*b*)). One then speaks of an array or a 'spectrum' of relaxation times. From this, a mean relaxation time can be calculated.

(*d*) *Normal stress component*

If a rod is rotated in a Newtonian liquid, the circular motion causes the formation of a vortex or funnel. If the same experiment is performed with a visco-elastic liquid the liquid may climb up the rod. This phenomenon, called the 'Weissenberg effect', has been noticed in polymer solutions, aged condensed milk, cake batter, eucalyptus honey, wheat flour dough, and perhaps most pronounced of all in malt extract (*see* Figure 10.8).

The Weissenberg effect is defined as a centripetal force observed during rotational shearing and may be visualized as follows: when a wire is stretched it becomes both longer and thinner. The opposite happens on compression, when it becomes shorter and thicker. If an iron cylinder is clamped firmly at the bottom and twisted at the top it does not normally become longer. Indeed it will break before a large strain occurs. If the experiment is repeated with a rubber cylinder or similar elastic material with a substantial twist applied,

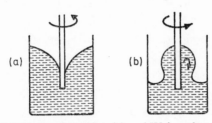

Figure 10.8 Vortex formation (*a*) and Weissenberg effect (*b*) on rotation of a rod in a Newtonian and a visco-elastic liquid respectively

it is found that the top surface is curved outwards. Hence there is a stress component at right angles to the shear force which is greatest at the centre of the specimen. In a visco-elastic liquid this results in the ascent of the liquid at the rod. This phenomenon was predicted on theoretical grounds by Karl Weissenberg and in his honour is referred to as the Weissenberg effect. It is readily measured by means of the Rheogoniometer[77] which is essentially a cone and plate viscometer where the upward force on the cone can be measured. In this way both the torsional and the vertical forces can be determined. From these results the behaviour of the visco-elastic material can be calculated.

(*e*) *Empirical*

Many solid foods are visco-elastic and very many empirical measurements on such foods have been described in the literature. Usually it is pretended that the material is either an elastic solid or a viscous liquid, and the other less important component is neglected altogether. For liquid measurements, recording dough mixers and other empirical viscometers have been used. For solid measurement, stretching, bending or compressing apparatus, penetrometers, and gel testers have been employed. Such measurements are frequently very useful in practice but have no theoretical basis.

10.5 SURFACE VISCO-ELASTICITY

Viscosity and elasticity are bulk properties. They represent the rheological behaviour of the bulk of the material. The rheology at the surface of a material may be very different. For example if lauryl alcohol is spread on the surface of water (the substrate), it can form a monomolecular layer, a layer one molecule thick, at the liquid–air interphase. The viscosity of this monolayer is about one million times greater than the bulk viscosity of lauryl alcohol. Many materials will give such monomolecular surface films. Surface active agents, detergents, proteins or non-aqueous solvents on water are examples. Liquid:liquid and other liquid:gas systems also have interfaces.

Surface viscosity has been defined as the change in viscosity of the surface layer of the substrate, brought about by the insoluble mono-layer. It is an empirical measure because of the drag of the layer immediately below the interphase. If the film is fairly substantial (protein films) rings, discs or needles suspended in the surface and oscillated are used to measure surface viscosity[78]. For thin films, a two dimensional capillary flow method is used. For a long narrow canal (*see* Figure 10.9) the equation is an analogy of the three dimensional Poiseuille flow (*see* Chapter 4):

$$Q = \frac{\pi d^3}{12\eta_s l}$$

Where Q is the volume of flow delivered in unit time (m³/s), π is the surface pressure of the film (N/m *not* N/m²), d and l are the diameter and length of the canal (m), and η_s is the surface viscosity (Ns/m, i.e. the units for bulk viscosity less one unit of length).

The value of η_s is often reported as apparent surface viscosity, because many surface films show non-Newtonian behaviour although the bulk of the liquid may show Newtonian properties.

Interfaces may also exhibit plastic or visco-elastic properties. Tachibana and Inokuchi[79] obtained loading–unloading curves on monolayers of various proteins spread over very dilute aqueous hydrochloric acid. They suspended a ring in the surface and loaded the system by twisting the suspension wire. The curves for protein

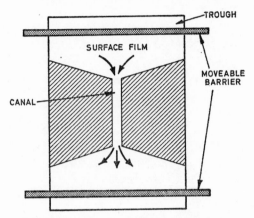

Figure 10.9 Canal type surface viscometer

and synthetic polypeptide films were similar to the deformation–time curve of the Burgers model. They showed viscosity and rapid and delayed elasticity.

11 Visco-elastic Materials: Examples

Materials exactly reproducing the Maxwell and Kelvin–Voigt models are not very common. Coagulated blood and cream approximate the former and stiff gelatine gels, the latter. On the other hand, the Burgers model provides a first approximation for very many foods. Starch-, gelatine- and milk gels, ice-cream, marshmallow, apples, cooked noodles, cheese, wheat flour dough, bread- and cake-crumb, protein foams, and wheat and maize grains all have been described as resembling it.

11.1 VISCO-ELASTIC LIQUIDS

All liquids exhibiting the Weissenberg effect (*see* Chapter 10) are visco-elastic. There are, however, many liquids which are visco-elastic but do not exhibit the Weissenberg effect.

Visco-elasticity in a liquid is easily demonstrated by boiling 1·5 g of starch in 100 ml of water to gelatinize it. The solution is then cooled and placed into a beaker. Another beaker is filled with 100 ml of sucrose solution. Both beakers are then slowly rotated, for instance by placing them on the turntable of a gramophone. If the rotation is suddenly stopped the inertia carries the syrup along a little at first but then it comes to rest. The same happens with the starch solution, except that once it has come to a complete stop, the flow is reversed and only after several oscillations does the solution finally come to a halt. The same experiment can be performed with a moderately dilute gelatine solution which is setting. Two other good examples of visco-elastic liquids are oleic acid, which has been dispersed in dilute ammonia and also 0·4 per cent aqueous polyethylene oxide.

An interesting experiment is that of Michaud[80]. He set up two flasks connected by a fine capillary (*see* Figure 11.1) and filled them with a very dilute solution of agar. After one or two days the liquid levels in the two bottles had equilibrated and were at the same height. If a glass rod was now immersed into either of the two flasks a tiny amount of liquid was displaced and the level rose in that flask by a

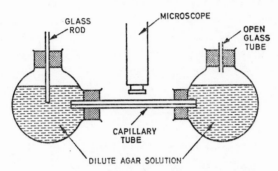

Figure 11.1 Demonstration of the solid characteristics of dilute agar solution[80]

minute amount. With a Newtonian liquid there would be immediate flow into the other flask to equilibrate the level. With the agar solution there was a slight resistance to flow. This was apparent if some carbon black had been dispersed in the liquid and the movement of the particles (or lack of it) in the capillary tube was observed by means of a microscope.

It appeared from these results, that very dilute agar solution, which usually behaves as a liquid, does not, in fact, resemble the Maxwell model (i.e. the visco–elastic liquid) but the Kelvin–Voigt model (i.e. the visco–elastic solid). The reason is that a finite force, even if small, is still required for flow to occur (*see* Chapter 10).

Michaud also discovered that this very slight rigidity disappeared, if the solution was heated above the temperature at which a concentrated agar gel normally melted.

This change from liquid sol to solid gel which occurs with starch, gelatin, agar, and many other hydrocolloids on heating, is important in canning because heat transfer is different with the two systems. In sols it is fast and occurs by convection in the tin. With gels it is slow and occurs by conduction. If a liquid starchy product is canned, heat transfer alters at about 70–80°C when the starch gelatinizes.

11.2 GELATINE GELS

Gels are usually hydrophyllic colloids possessing a three dimensional network of molecules. In relatively short time experiments, gelatine gels can be made to behave as a solid showing rubber-like elasticity (*see* Chapter 2). A convenient method for measuring the shear or rigidity modulus of gels is due to Saunders and Ward[81]. The gel is allowed to set at a controlled temperature in a tube which ends in a capillary U-tube filled with mercury (*see* Figure 11.2). Air pressure

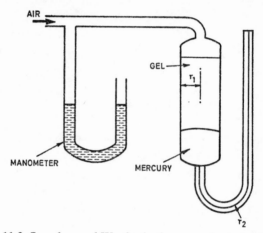

Figure 11.2 Saunders and Ward tube for measuring gel rigidity[81]

is applied to the other opening of the reservoir and the volume deformation caused by the air pressure is measured from the rise of the mercury in the capillary.

In this experiment the shear or rigidity modulus G equals

$$\frac{Pr_1{}^4}{8lr_2{}^2h}$$

where P is the air pressure less the back pressure of the mercury, r_1 is the reservoir radius, r_2 is the capillary radius, l is the height of the gel, and h is the displacement of mercury in the capillary.

For soft gels, a wide U-tube has been used with coloured carbon tetrachloride in the capillary and kerosene in the manometer. Gelatine gels were found to be almost, if not quite, Hookean. The shear modulus depended very slightly on the strain. It was shown to be of the order of 6600–6800 N/m² for a 5·5 per cent aqueous gel at 10°C. Using the mercury bath extensometer, similar to that used for flour doughs, a Young's modulus of the order of 600–2000 N/m² was found for various gelatine gels (10 per cent concentration at 10°C). In these experiments the viscous deformation was about 1 per cent[82].

In experiments of longer duration the viscous component becomes much more pronounced. In one experiment[83] a concentric cylinder viscometer was used. The gel was placed into the gap and the outer cylinder rotated through a small angle. The inner cylinder held by a stiff suspension responded, and its rotation over 20 hours allowed the relaxation time T_M to be determined. In this way it was shown that relaxation was due to breaking of bonds which subsequently reformed in the unstrained configuration.

11.3 WHEAT FLOUR DOUGH

Dough has been described as 'the most complicated rheological material at present known to man' (M. Reiner). Because the rheological properties of dough are closely connected with baking characteristics of flour, tests had to be developed long before visco-elastic measurement was properly understood. Hence, these tests were empirical and are still used to-day on a very large scale because nothing better is available[8].

(a) Empirical tests

Three dough testing systems are in commercial use: The Chopin Alveograph (mainly in France), the Simon Extensometer (mainly in U.K.) and the Brabender system (world-wide).

The Brabender system comprises firstly a recording dough mixer, the Farinograph[84]. The flour is placed into the mixing bowl and water run into it from a burette. As the dough is being mixed at 30°C, the consistency is automatically recorded on a chart. When a consistency of 500 empirical units is indicated, the mixer is stopped and the water added from the burette gives the water absorption of the flour. The test is now repeated but this time the water determined is added to the flour at once, and the mixing curve obtained is characteristic of the flour. Figure 11.3 shows the curve of a strong and a soft flour. The former contains about 12 per cent of protein and is suitable for bread baking. The latter contains about 8 per cent of protein and is suitable for cake or biscuit production.

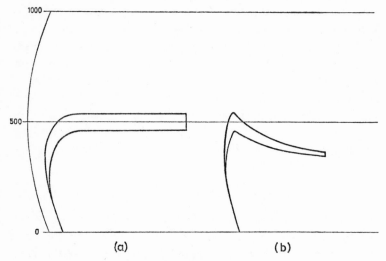

Figure 11.3 Farinograms of: (*a*) strong flour dough;
(*b*) soft flour dough

The second instrument of the Brabender system is the Extenso-graph. This instrument forms a dough cylinder containing some salt and a standard amount of water (previously determined with a Farino-graph) and after 45 min resting time at 30°C, stretches the dough until it tears. A load extension curve is obtained which is again typical of the flour used (*see* Figure 11.4).

Farinograms and Extensograms have been referred to as 'finger prints' of flour. They have little fundamental meaning but allow reasonable classification of flours for baking purposes. Attempts have been made to use the Extensograph for fundamental work but the methods are complicated and only useful for research, not plant control work.

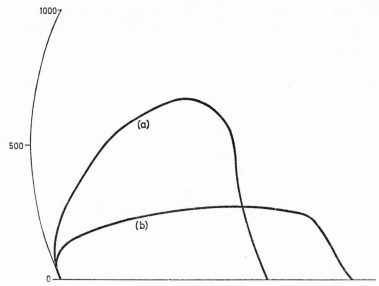

Figure 11.4 Extensograms of: (*a*) strong flour dough;
(*b*) soft flour dough

(*b*) Fundamental tests

All of the four different techniques for visco-elastic measurement given in Chapter 10 have been used with dough. Measurement of the normal stress component is unsuccessful because the dough disc between the cone and plate rolls into a cylinder[85]. The other three approaches (vibrational[85], loading–unloading[86], relaxation[87]) work well, but are not used industrially because they are laborious and have consequently not been correlated with baking performance. (With empirical tests, we do not know what we are doing, but it works; with fundamental tests we know what we are doing, but it doesn't!) Figure 11.5 shows the deformation time curve for dough at different stresses using the mercury bath extensometer. From these results, viscous, and rapid, delayed and total elastic deformation can be obtained and plotted against tensile stress (Figures 11.6 and 11.7).

What rheological model can we assign to dough? It is apparent from Figure 11.5 that dough possesses a rapid elastic deformation and therefore it contains a free Hooke element. It also possesses a residual irrecoverable deformation (*see* Figure 11.5) and hence a free Newton element. It has already been shown that a Hooke and a Newton element in series are referred to as a Maxwell model, which when stretched to a fixed deformation, exhibits the phenomenon of stress relaxation. Dough shows this behaviour, but it is apparent that

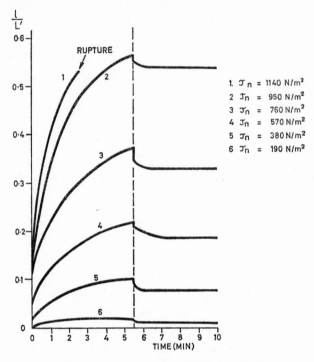

Figure 11.5 Creep and recovery experiment: Deformation versus Time at different constant stresses

it also shows delayed elastic deformation (*see* Figure 11.5). Hence there must be a Kelvin–Voigt component. Since a Maxwell and Kelvin–Voigt model in series are referred to as a Burgers model, it is apparent that dough closely resembles this.

Although this model is quite adequate for most practical purposes, there are dough properties which the dough model does not reflect[86]. Since there is no recoverable deformation below a certain minimum stress (a yield point), a St. Venant element must be placed in parallel with the Hooke element of the Maxwell part. Since in relaxation experiments there always remains a residual stress, there must also be a St. Venant element in parallel with the Newton element of the Maxwell part.

Many dough parameters are not linear (*see* Section 10.4(*b*)). Non-linearity cannot be conveniently depicted using spring and dash pot

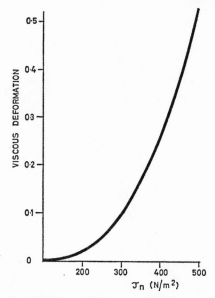

Figure 11.6 Irrecoverable or viscous deformation after specific time of loading versus tensile stress

models because they and their combinations (Maxwell, Kelvin–Voigt, Burgers models, Maxwell elements in parallel, Kelvin–Voigt elements in series, etc.) are all linear. It is possible to use 'non-Newtonian dash pots' or 'non-Hookean springs' but this is not very helpful and not normally done. So we may say if for example a Hooke element (*H*) is non-linear with respect to stress that it is modified by a function f, of stress or $H\,f(\tau_n)$. When the immediate elastic deformation is plotted against stress, the relationship is S-shaped (sigmoid). Hence $H_2\,f(\tau_n)$. In recovery and in viscometer experiments, the viscosity depends both on stress (by a power equation) and on time, hence $N_2\,f(\tau_n, t)$. The residual stress in relaxation experiments depends on the

5

original extension, hence $St.V_2$ f(e). The plot of delayed elasticity versus stress is sigmoid, hence H_1 f(τ_n).

Finally it has been established in recovery experiments that if a certain stress is exceeded, H_1 and H_2 decay. This is indicated by a so-called transformation element, T.

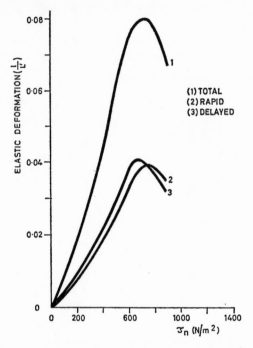

Figure 11.7 Rapid, delayed, and total elastic deformation versus stress

When the shear pin (*SP*) breaks (*see* Figure 11.8), the stress is thrown from the Hooke to the Newton element. The model for dough so far established is shown in Figure 11.9. It shows twelve rheological components.

The viscous and elastic constants for dough are very variable. They depend to a very large extent on the water content of the dough and even for the same dough the apparent coefficient of viscosity may vary about a thousand-fold depending on test conditions. The bulk modulus of bread dough has been given as 1×10^7, the shear

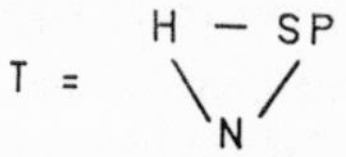

$$T = \begin{array}{c} H \ - \ SP \\ \diagdown \ \diagup \\ N \end{array}$$

Figure 11.8 Transformation element

or rigidity modulus as $1\text{--}10 \times 10^4$ and the apparent viscosity as $1 \times 10^4\text{--}10^7$ SI units.

As with gelatine, dough or gluten experiments can be designed, where the viscous deformation becomes very small. If that is so, the

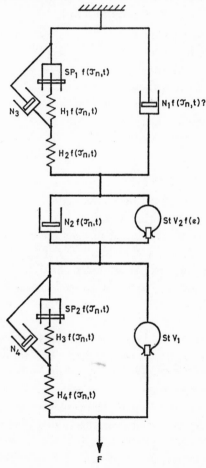

Figure 11.9 Rheological model of wheat flour dough[86]

rubber elasticity theory (*see* Chapter 2) can be applied which allows the molecular weight between cross links to be determined. It was shown that a strong flour gluten is more heavily cross linked than a soft flour gluten[16]. On heating, denaturation of the gluten occurs and this is expressed as an increase in cross linkage[88]. Flour improvers seem to act in the same way (*see* Table 11.1).

TABLE 11.1

Average molecular weight of gluten chain between cross links[16]

Strong gluten	$8{\cdot}9 \times 10^6$
Soft gluten	$10{\cdot}9 \times 10^6$
Strong gluten	$9{\cdot}3 \times 10^6$
Above with improver (100 ppm KIO_3)	$7{\cdot}4 \times 10^6$
Strong gluten	$8{\cdot}4 \times 10^6$
Above heated to 75°C	$3{\cdot}2 \times 10^6$

11.4 BAKED PRODUCTS

The baked products most generally studied are cake, bread, and biscuits. Cake consists virtually entirely of crumb, bread of an inner crumb and an outer crust, and biscuits almost entirely of crust. Crumb has been defined as the soft interior of a cake or loaf of bread. It has also been defined as air holes glued together by vitamin enriched edible plastic. As such we would expect it to have visco-elastic properties, and creep curves have in fact been determined for Madeira cake[89].

It was found that both the Young's modulus and apparent coefficient of viscosity increased with staling. The former from about 2×10^4 to 20×10^4 N/m². The apparent coefficient of viscosity increased about ten times. A compressimeter for bread crumb is shown in Figure 11.10. For a test on bread using this instrument, the loaf is cut into slices of a given thickness and a plastic square placed on each slice so that a square test sample can be cut out easily. The crumb is compressed for a certain time and then the weight removed. Both compression and decompression are determined from the pointer position on the scale. Young's modulus for bread crumb is of the order of about 2×10^4 N/m² but flour strength, fermentation time, baking time, and storage temperature affect it.

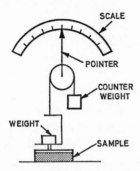

Figure 11.10 Compressimeter for bread crumb

Some empirical measure of biscuit hardness has been obtained by using a small disc saw to cut through a stack of biscuits[90]. The time taken for a standard cut to be made is recorded. The longer the time, the harder is the biscuit. Table 11.2 gives some results for various biscuits.

TABLE 11.2

Empirical hardness of biscuits[90]

Butter puff	8 s	Water biscuits	49 s
Cream cracker	13 s	Ginger nut	88 s
Shortbread	24 s		

Although this instrument shows quite clearly the hardness of different types of biscuits, it is not sensitive enough to give an accurate record of the variations in hardness of the same type of biscuit.

12 Solids in Contact

In some respects writing a book is like a love affair. Any fool can start one, but to end it requires considerable skill.

Perhaps a book on food rheology is not so difficult to complete because there remains one topic of great rheological interest which is often neglected, and that is the flow of solid particles. These can be large, like apples or oranges, or very small like particles of flour, sugar, milk or egg powder.

The flow of such particulate solids is of great industrial importance. In enters into the design of spouts, shutes and pneumatic conveying systems. It is as important in the construction of storage bins holding hundreds of tons of flour as in the design of the small containers in coffee vending machines which deliver a few grams of instant coffee at the drop of a coin.

12.1 UNIT AND BULK DENSITY

An important characteristic of particulate solids, whether they be of small size like powders, or of large size like fruit, is their density. First there is the unit density sometimes referred to as the 'real' density. This is the average mass per unit volume of the individual particles. It is determined by weighing the particle in air and determining its volume by displacement of a liquid, usually water. The ratio of weight (kg) divided by volume (m^3) is the unit density. If the material is of small particle size, a gradient tube is employed. It is filled with two miscible liquids of different densities and allowed to equilibrate for several days. Glass beads of known densities are immersed in the liquid and from their height at constant temperature, a graph of density versus height in the tube is constructed. The specimen is then introduced and from its height the density can be determined using the calibration curve. Table 12.1 shows some unit densities.

Secondly, there is the bulk or 'apparent' density. This is the mass per unit volume of the particles in bulk. The bulk density is always much lower than the unit density because of the large number of spaces between the particles. Table 12.1 shows some bulk densities. They are determined in the same way as unit densities, except that the material is placed in bulk into a plastic bag. Sand or seed displacement has also been used. For instance the standard method of determining the volume of a loaf of bread is by displacement of rape seed which is small and spherical and so packs evenly and well.

TABLE 12.1

Various unit and bulk densities (kg/m³)

	Unit density	*Bulk density*
Apples	700 – 900	545 – 608
Oranges	930 – 950	768
Cherries	970 – 1050	720
Potato	1120 – 1150	768
Maize	1220 – 1240	675
Rice	1360	575
Barley	1420	560 – 640
Wheat	1420	800
Wheat flour	1450	450 – 550

12.2 FRICTION

Friction is the name given to forces offering resistance to relative motion between solid surfaces in contact. The coefficient of friction is the force of friction parallel to the surface, divided by the normal force applied between the surfaces. It is as a first approximation independent of the area of contact. If the two bodies in contact are made to slide over one another from rest, friction is relatively high (static friction). Once they slide, a slightly smaller force will maintain the motion (kinetic friction).

The magnitude of the two coefficients of friction (the static and the kinetic coefficients) depend on the materials in contact, the types of surface, the temperature, and the velocity of movement.

A common method of measuring friction of relatively large objects like grain or fruit is to place them into a square frame on a rotating disc[91]. The friction is readily determined from a scale (*see* Figure 12.1).

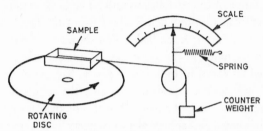

Figure 12.1 Rotating disc method for determination of friction

Some coefficients of friction are given in Table 12.2.

TABLE 12.2

Some coefficients of friction

	Static	*Kinetic*
Tomatoes on metal	0·36	0·31
Tomatoes on plastic foam	0·74	0·71
Apples on metal	0·40	0·31
Apples on plastic foam	0·72	0·60

12.3 ANGLE OF REPOSE

Both liquids and particulate solids flow. If we pour some liquid into a beaker, we will find that the liquid comes to rest with a free horizontal surface (if we neglect the slight curvature due to surface tension). If we repeat the experiment with a powder, say flour, we find that the material comes to rest in a conical heap. The angle of the side of the heap to the horizontal differs with different solids. It is referred to as the 'angle of internal friction' or the 'angle of repose' (*see* Figure 12.2). Usually the drained angle of repose (*see* Figure 12.2(*a*)) is the same as the poured angle of repose (*see* Figure 12.2(*b*)).

Any particle outside the slope will slide down. Within the surface and below it, the particles are in equilibrium, the gravitational pull being balanced by particle interaction. So the angle of repose has a similar significance to the yield value with plastic materials (*see*

Chapter 8): it divides two types of rheological behaviour. With free flowing powders it is possible to speak of 'powder viscosity', with settled powders of 'powder plasticity'. Just as the yield value of a plastic material can be made to disappear by addition of a 'liquid-like' phase (*see* Chapter 8), so the angle of repose and the plastic properties

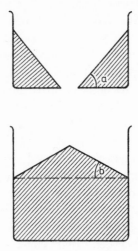

Figure 12.2 Drained (*a*) and poured (*b*) angles of repose

of a settled powder can be obviated by vibration or fluidization (*see* Section 12.5).

Table 12.3 shows the angle of repose of various foodstuffs. It is affected by density, particle size and distribution, and the surface characteristics of the particles.

TABLE 12.3

Angle of repose of various granular foodstuffs (degrees)

Rice	20	Granulated sugar	35
Maize	21	Cocoa powder	40
Wheat	23	Self-raising flour	41
Barley	30	Flour	45
Granulated salt	31	Icing sugar	45

12.4 POWDER-VISCOSITY

Although the use of egg timers is very ancient, and the observation of sandstorms even more so, the first systematic study of powder flow was conducted by Bingham and Wikoff[92]. These workers used sand but the same results are obtainable with fine semolina. The semolina is placed into a small reservoir terminating in a capillary tube at the bottom. Bingham and Wikoff found that while increasing the temperature of a liquid decreased its viscosity, temperature effects on powders were negligible. When the head of a liquid is increased, flow increases. With powders flow decreased, probably due to compaction. When the length of the capillary tube was increased, flow of liquid decreased, but the flow of powder increased, presumably because the increased length of the capillary decreased turbulence. With powders the flow rate increased with decreasing particle size, while with liquids the flow rate increased with decreasing molecular size. Recently, experiments on powders have been conducted using the equivalent of an orifice viscometer. The results obtained were similar to those obtained with the capillary tube except that it was found that the head of powder was independent of flow rate.

Using sugar, milk powder, instant coffee, and drinking chocolate it was found that the mass flow rate (kg/s) through the orifice was roughly proportional to the bulk density and to the 2·8th power of the orifice diameter (*see* Figure 12.3). This rule is only applicable if the material is free flowing and does not bridge or form agglomerates.

Just as particle interaction in solutions is of great importance to liquid viscosity, so the stickiness of powders is of importance in powder viscosity.

Powder stickiness is not well understood but it is known to be a surface property. It causes many industrial problems on storage and in conveying. Often a sticky powder can be converted into a free flowing one by drying or fat extraction. Flour and starch are examples of sticky powders. When extracted with ether and dried, they become free flowing and are used industrially for dusting purposes. These products are characteristically dusty to handle, have a lower bulk density and a large amount will pass through a screen for a given amount of shaking.

Stickiness may also be assessed by adding sufficient fine sand to the sticky powder for the mixture to maintain a steady flow through a small hole in the container.

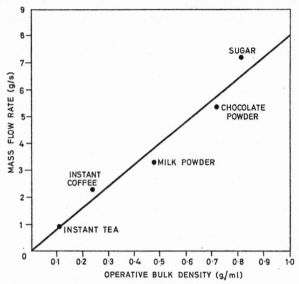

Figure 12.3 Dependence of mass flow rate on bulk density[93]

12.5 FLUIDIZATION

Normally, powders are two phase systems of solid and gas, generally air. If powder is placed on a porous surface, and air blown through it, several important changes occur. At very low air speeds there is simply a percolation of the air through the solid. If the air flow is increased, the mass expands, forms a flat horizontal surface much like a liquid, the angle of repose disappears, and low density materials will float on the aerated powder as if on water. This process has been referred to as a 'fluidization' and the experimental arrangement as a 'fluidized bed'[94]. So fluidization of a powder is similar to adding 'liquid-like' phase to a plastic material. It results in greater mobility. The velocity of the air required to fluidize the particles is called the 'minimum fluidizing velocity'.

If the air flow is further increased, bubbles will form and carry excess air with them. If the flow is increased further still, air and powder will form a mixture very similar to a dilute suspension.

Matheson *et al.*[95] have studied the powder viscosity of fluidized beds. They measured the torque required to rotate the paddle of a viscometer at 200 r.p.m. in a 3-in. bed of solids fluidized with air (*see* Figure 12.4). Powder viscosity depended on the type of powder, the

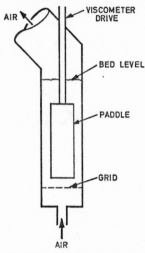

Figure 12.4 Paddle viscometer[95]

size and size range, the evenness of size, and the tendency to spherical shape. It also depends on the velocity of the fluidizing air, increasing rapidly near the minimum fluidizing velocity.

The movement of the particles in the air-stream allows very rapid heat transfer between the two phases. For this reason fluidization has found useful applications in the freezing of particulate foods such as peas. Similarly, 'floating' cans of foodstuffs in a fluidized bed of heated particles may offer an alternative to conventional retorting of cans in steam.

Powders such as flour or sugar can be transported in 'dilute suspensions' through pipelines. This application of fluidization can substantially reduce the costs of bulk handling.

Appendix A:
Laboratory Experiments

1. MEASUREMENT OF THE DENSITY OF A LIQUID

The density of a liquid is required for the determination of the coefficient of viscosity (*see* Chapter 4).

Weigh a clean 50 ml ($50 \times 10^{-6} m^3$) density bottle on an accurate analytical balance. Fill it with the liquid under test and equilibrate to correct temperature. Dry the bottle carefully on the outside and weigh. Subtraction of the weight of the empty bottle from this figure gives the weight of its content. Division of this weight in kg by the volume of the bottle in m^3 gives the density of the liquid in kg/m^3.

Repeat the test and report the mean of the two readings.

2. DETERMINATION OF THE YOUNG'S MODULUS OF DRY SPAGHETTI

Clamp two scalpels rigidly on to two retort stands so that the blades are vertical. Adjust the blades so that they are 0·200 m apart and about 0·15 m from the surface of the table. Place one stick of dry spaghetti across the upper knife edges. If possible, the spaghetti should have been equilibrated at 25°C and 65 per cent R.H. (over a concentrated solution of sodium nitrate) for 48 hours.

Apply several successive loads (kg) at the centre of the spaghetti and measure the vertical deflection d (m). The deflection for each load may be measured with the aid of an accurately machined steel ruler and hand lens, or preferably a travelling microscope. The loads may be placed on a small balance pan made from tin foil and fine wire. Plot the weight applied against the deflection for each weight. Draw a straight line through the points. Determine the deflection from any convenient point on the line and the respective weight applied.

Young's modulus E equals:

$$\frac{MgL^3}{12\pi r^4 a} \; N/m^2$$

where M is the weight applied (kg), g is the gravitational constant (9.81 m/s^2), L is the distance between the scalpel blades, π is 3.1416, r is the radius of the spaghetti (m), and a is the deflection (m) at the centre of the spaghetti (*see* Figure 2.4 (*c*)).

3. MEASUREMENT OF THE COEFFICIENT OF VISCOSITY OF A NEWTONIAN LIQUID, USING THE OSTWALD CAPILLARY VISCOMETER

The Ostwald viscometer is shown in Figure 4.8. Pipette the correct volume of distilled water into arm B of a clean viscometer. Immerse it vertically into a large beaker with water or a thermostatically controlled bath. Bring to constant temperature ($25°C$ may be convenient). Draw up the water above mark C in arm A and time its fall from mark C to mark D. Repeat twice. Discard the water and dry the instrument. Repeat the procedure using the test liquid.

If η_1 is the viscosity of water at the temperature used (*see* Appendix B for this value), t is the time in s for the water to fall from mark C to mark D, ρ_1 is the density of the water in kg/m^3 at that temperature (*see* Appendix B for this value) and ρ_2, t_2, and η_2 refer to the test liquid, η_2 in Pl is obtained from the equation:

$$\frac{\eta_1}{\eta_2} = \frac{t_1}{t_2} \times \frac{\rho_1}{\rho_2}$$

As an example milk may be used for this experiment.

4. MEASUREMENT OF THE COEFFICIENT OF VISCOSITY OF A NEWTONIAN LIQUID USING THE FALLING BALL METHOD

The apparatus is shown in Figure 4.9. Fill a measuring cylinder with the liquid under test and bring to the required temperature. $25°C$ is suitable. Mark the cylinder at A and B. Mark A should be well below the level of the liquid so that the ball (a steel ball bearing is convenient) has reached a steady velocity when passing the mark. Determine the fall time in s of the sphere between A and B. The density of steel is 7830 kg/m³. Determine the density of the liquid using the method given in Appendix A, Section 1.

$$\eta = \frac{2}{9} g(\rho_1 - \rho_2) \times \left(\frac{a}{b}\right)$$

where η is the viscosity of the liquid in Pl, g is the gravitational constant (9·81 m/s²), ρ_1 is the density of the sphere in kg/m³, ρ_2 is the density of the liquid and a/b is the terminal velocity of the sphere (a is the distance $A - B$ in m, b is the fall time in s).

As an example a 60 per cent solution of sucrose may be used.

5. MEASUREMENT OF THE APPARENT VISCOSITY OF A NON-NEWTONIAN LIQUID USING A NARROW GAP ROTATING CYLINDER VISCOMETER

For this experiment the Haake Rotovisco instrument is used. This is a narrow gap concentric cylinder viscometer and is electrically driven. The temperature is kept constant at 25°C. Speeds available are 3–486 r.p.m. and the viscosity of the liquid is determined from a scale reading 0–100 units. The liquid tested may be cream of tomato soup. The soup is gently poured into a beaker to break up the gel and from there into the sample cup. The sample is left for 300 s to reach the required temperature. The instrument is then run on the highest gear number for 30 s. The scale reading is determined. The instrument is switched off for 5 s during which time the next gear number is set. The instrument is then run for a further 30 s, switched off for 5 s and the process repeated until the lowest gear number (highest speed) is reached. The speeds are then decreased in the same way.

From the values of gear number (speed) and scale reading, τ and D are determined from the calibration charts provided by the manufacturer. The curves of τ versus D and η_{app} versus D are plotted (*see* Figure 7.4).

6. MEASUREMENT OF THE HESION VALUE OF A PLASTIC FOOD

The apparatus (*see* Figure A.6.1) consists of a metal rod D, which is suspended at the centre to act as a balance beam. The rod carries at one end a light container E (plastic ice cream tub), and at the other a counter weight C, and a square piece of plastic B. The material to be tested is placed into beaker A and the plastic inserted vertically to the mark. A convenient size for the plastic below the mark is 0.025 m $\times 0.025$ m. Water is run into container E from a burette F. When

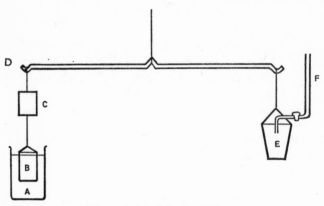

Figure A.6.1 A hesion balance

the hesion value is reached the plastic square is torn out of the test material.

$$\text{Hesion value} = \frac{a \times \rho \times g}{2A} \text{ N/m}^2$$

where a is the volume of water in litres, ρ is the density of the water in kg/m³ (*see* Appendix B), g is 9.81 m/s², and A is the area (m²) of the plastic sheet. (*Note :* the area is multiplied by 2 since two sides are in contact with the material).

As an example mashed potato may be used.

7. EMPIRICAL MEASUREMENT OF THE HARDNESS OF FAT

Place the fat into a beaker and melt slowly in a water bath. Do not overheat the fat. Pour the melted fat into a box about 0.07×0.07 m in width and 0.03 m deep. Cool in a refrigerator overnight. Place into a thermostatically controlled water bath set at say 20°C for

2 hours. Drop a ball bearing of 0·035–0·040 m diameter on to the surface of the fat from a height not exceeding 0·075 m.

Before dropping, the sphere may be held at the correct distance by an electromagnet or a collar fitted with two fixed screws and a wingnut.

Where the fat touches the bright surface of the sphere it leaves a dull mark, the diameter of which is readily determined using dividers and an accurately machined steel ruler. The diameter of the mark is inversely proportional to the hardness of the fat.

8. MEASUREMENT OF THE APPARENT COEFFICIENT OF VISCOSITY AND SHEAR MODULUS OF WHEAT FLOUR DOUGH USING THE MERCURY BATH EXTENSOMETER

The apparatus is shown in Figure 10.5(*a*). A dough is mixed and transferred to an extruder with a nozzle of about 5 mm diameter. A water-jacketed grease gun can be used for this purpose. The dough extruded during the first 15 s is discarded. A portion of the dough extruded subsequently is cut off by a special hinged knife (*see* Figure 10.5(*b*)). Simultaneously two strips of adhesive tape are wrapped around the ends of the dough cylinder to form cuffs. One cuff carries a thread for attachment to the rigid support of the extensometer, the other carries a similar thread for attaching to the movable arm. The dough cylinder is now coated with olive oil to reduce friction on the mercury and decrease evaporation and is now ready for the test. On a separate run the length of the dough is measured, and its volume determined by displacement of kerosene. The volume is divided by the dough length to give the average cross-sectional area which is later required to convert the load into stress (stress equals load per unit area of cross-section). This is necessary because the cross-sectional area of the dough cylinder is larger than the extruder nozzle, the increase being dependent on dough properties. Several dough cylinders are then extended under loads of 2 to 6 g, one cylinder for each load. (This load must be corrected for the friction between the dough cylinder and the mercury. The load—usually of the order of 1 g—which will just cause an untethered dough cylinder to move on the mercury surface is determined at the start of each experimental series. Since during extension one end of the dough is fixed, half the value thus measured is subtracted from the experimental load.) After 5·5 min the load is simply removed by burning the thread connecting the dough and the balance pan. The dough lengths, during both extension and recovery, are measured to the nearest mm using a steel ruler. If the original length of the dough piece is L and each increment measured is l, then L' at any time of the experiment equals $L + l$.

The deformation is l/L'. The stress is equal to $(W - F/2)/A$ where W is the load (to compensate for the leverage of the movable arm), F is the friction, and A is the cross-sectional area of the dough at the start of the experiment. Figure 11.5 shows the plot of l/L' against time for various loads.

Now viscous, rapid elastic and delayed elastic deformations (*see* Figure 11.5) determined for various loads on unloading may be plotted against stress (*see* Figures 11.6 and 11.7). The average apparent coefficient of viscous traction now equals shear stress divided by the average rate of viscous deformation (viscous deformation divided by time of loading).

Similarly, when total, rapid and delayed elastic deformation is divided by the stress the three respective moduli are obtained. Of these only the rapid elastic modulus approaches a constant value.

Appendix B: Viscosity and density of water

°C	Viscosity (Pl)	Density (kg/m³)
10	0·0013060	999·7
15	0·0011380	999·1
20	0·0010020	998·2
25	0·0008901	997·1
30	0·0007974	995·7

Appendix C: Symbols, Units, and Conversion Factors

1. SOME SYMBOLS USED

F	Force
A	Area
τ	Shear stress
τ_n	Tensile compressive stress
τ_v	Hydrostatic pressure
γ	Displacement gradient
μ	Poisson's ratio
V	Volume, sphere velocity
e	Longitudinal strain, annular deflection of cylinder
G	Shear modulus
E	Young's modulus
K	Bulk modulus, instrumental constant
e_v	Volumetric strain
D	Rate of shear
η	Coefficient of dynamic viscosity
η_{app}	Apparent coefficient of viscosity
η_{pl}	Plastic viscosity
Pl	Poiseuille
H	Hooke model
N	Newton model
$St.V$	St. Venant model
ρ	Density
R_c	Critical Reynolds Number
Q	Volume delivered at end of tube
r	Tube, cylinder radius

l	Length of tube, cylinder
t	Time (s)
p	Pressure difference
g	Gravitational constant ($9 \cdot 81$ m/s^2)
h	Height
ω	Angular velocity
R	Mean cylinder radius, gas constant
W	Weight
log	Log base 10
ln	Natural log
TS	Total solids
BP	Boiling point
pH	Hydrogen ion concentration
τ_0	Yield value
Bi	Bingham model
π	$3 \cdot 1416$

2. UNITS AND THE SI SYSTEM

Thirty countries, including the United Kingdom, have decided to adopt the SI (Système International) system of units. This system has been used in this book. However, much of the older literature gives other systems of measurement and for this reason the following units and conversion factors are given.

The six basic SI units

Physical quantity	*Name of unit*	
Length	metre	(m)
Mass	kilogram	(kg)
Time	second	(s)
Electric current	ampere	(A)
Thermodynamic temperature	degree Kelvin	(°K)
Luminous intensity	candela	(cd)

Derived or allowed units used in this book

Physical quantity	Name of unit	Definition of unit
area	square metre	m^2
volume	cubic metre	m^3
density	kg per m^3	kg/m^3
velocity	m per s	m/s
acceleration	m per s^2	m/s^2
energy	joule (J)	$kg\ m^2/s^2$
force	newton (N)	$kg\ m/s^2$
power	watt (W)	$kg\ m^2/m^3$
pressure, shear stress	N per m^2	N/m^2
dynamic viscosity	poiseuille (Pl)	$N\ s/m^2$
customary temperature	degree Celsius (°C)	$T(°K) - 273·15$
litre	litre (l.)	$10^{-3}m^3$

Conversion Factors

To convert from	to	multiply by
centimetres	inches	0·3937
inches	centimetres	2·54
metres	feet	3·2808
feet	metres	0·3048
ounces	grams	28·35
grams	ounces	0·03215
kilograms	pounds	2·2046
pounds	kilograms	0·45359
litres	pints	1·7598
pints	litres	0·5682
litres	gallons	0·21998
gallons	litres	4·54596
psi	dynes/cm²	$6·895 \times 10^4$
dynes/cm²	psi	$1·450 \times 10^{-5}$
psi	kg/mm²	$7·030 \times 10^{-4}$
kg/mm²	psi	$1·422 \times 10^3$

General Bibliography

(a) GENERAL AND THEORETICAL RHEOLOGY

Reiner M. *Deformation, Strain and Flow*. Lewis, 1960. (Standard general text)

Reiner, M. *Lectures on Theoretical Rheology*. Interscience, 1960. (Theoretical and advanced)

Jaeger, J. C. *Elasticity, Fracture and Flow*. Methuen, 1964. (Theoretical and advanced)

Eirich, F. R. *Rheology: Theory and Application*. Academic Press. Vol. 1, 1956; Vol. 2, 1958; Vol. 3, 1960; Vol. 4, 1968. (Large reference work on general rheology)

Van Wazer, J. R., Lyons, J. W., Kim, K. Y., and Colwell, R. E. *Viscosity and Flow Measurement*. Interscience, 1963. (Valuable reference book on viscometers. Theoretical chapters advanced)

Scott-Blair, G. W. *Elementary Rheology*. Academic Press, 1969.

(b) RHEOLOGY OF FOOD

Sherman, P. *Industrial Rheology*. Academic Press, 1970. (At present the only advanced and comprehensive book on Food Rheology)

Mohsenin, N. N. *Physical properties of plant and animal materials*. Gordon and Breach, Vol. 1. 1970. (Comprehensive work containing sections on food rheology)

Scott-Blair, G. W. (Ed.) *Food Stuffs, their plasticity, Fluidity and Consistency*. North Holland, 1953. (Mainly of historical interest)

Society of Chemical Industry. *Monograph 7. Texture in Foods*. London, 1960. (This and the following book contain chapters by various authors. Some are good, some are not)

ibid. *Monograph 27. Rheology and Texture of Foodstuffs*. London, 1968.

References

1. Muller, H. G. *J. Texture Studies.* 1, 38, 1969
2. Scott-Blair, G. W. *Elementary Rheology.* Academic Press, 1969
3. Reiner, M. *Deformation, Strain and Flow.* Lewis, London, 1960
4. *Judges* 5, 4–5: Song of Deborah and Barak
5. Poggendorf, J. C. *Biographisch–Literarisches Handwörterbuch.* Akademie Verlag Berlin, 1863–1970
6. Reiner, M. *Lectures on theoretical rheology.* Interscience, 1960
7. Jaeger, J. C. *Elasticity, Fracture and Flow.* Methuen, 1964
8. Muller, H. G. *J. Fd. Technol.* 4, 83, 1969
9. Mohsenin, N. N. and Morrow, C. T. *Society of Chemical Industry Monograph 27*, p. 50, 1968
10. Finney, E. E. and Norris, K. H. *60th Ann. Meeting Amer. Soc. Agric. Eng. Saskatoon, Saskatchewan,* 1967
11. Epstein, B. *J. Appl. Phys.* 19, 140, 1948
12. Muller, H. G., Williams, M. V., Russell-Eggitt, P. W., and Coppock, J. B. M. *J. Sci. Fd. Agric.* 12, 513, 1961
13. O'Neill, H. *The Hardness of Metals and its measurement.* Chapman and Hall, London, 1934
14. Treloar, L. R. G. *The physics of rubber elasticity.* Clarendon Press, Oxford, 1958
15. Saunders, P. R. and Ward, A. G. in Mason, P. and Wookey, N. *Rheology of Elastomers.* Pergamon, p. 45, 1958
16. Muller, H. G. *Cereal Chem.* 46, 443, 1969
17. Wiederhorn, N. M. and Reardon, G. V. *J. Polymer Sci.* 9, 315, 1952
18. Cater, C. W. *J.S.L.T.C.* 49, 455, 1965
19. Karacsonyi, L. P. and Borsos, A. C. *Cereal Chem.* 38, 14, 1961
20. Cray, R. E. *Poult. Process Market.* 59, 10, 1953
21. Brooks, J. *Society of Chemical Industry Monograph 7*, p. 149, 1960
22. Swenson, T. L. and James, L. H. *U.S. Egg Poult. Mag.* 38, 14, 1932
23. Sluka, S. T., Besch, E. L., and Smith, A. H. *Poultry Sci.* 24, 1494, 1965
24. Wolodkewitch, N. N. *Z. Lebensm. Unters. u. Forsch.* 105, 1, 1957
25. L.E.E. Inc. 625 New York Ave N.W. Washington, D.C.
26. Instron Engineering Corp. 2500 Washington Ave, Canton, Mass.
27. Proctor, B. E., Davison, S., Malecki, G. J., and Welsh, M. *Fd. Technol.* 9, 471, 1955

28. Friedman, H. H., Whitney, J. E., and Szczesniak, A. S. *J. Fd. Sci.* 28, 390, 1963
29. Newton, I. S. *Philosophiae Naturalis.* Book 2, Section 9, 1687
30. Ward, A. G. *Trans. Faraday Soc.* 33, 88, 1937
31. V. Wazer, J. R., Lyons, J. W., Kim, K. Y., and Calwell, R. E. *Viscometry and Flow Measurement.* Interscience, 1963
32. Onyon, P. F. in Allen, P. W. (Ed.) *Techniques of Polymer characterisation.* Ch. 6. *Viscometry.* Butterworth, 1959
33. *Determination of Viscosity of Liquids in Absolute Units. British Standards* 188, 1937, (Inc. 1951 Amendments)
34. Merrington, A. C. *Viscometry.* Arnold, 1949
35. Sherman, P. *Industrial Rheology.* Academic Press, 1970
36. Bates, F. J. *et al. Polarimetry and Saccharimetry.* U.S. Bureau of Standards Circular C440, 1942.
37. Lipscombe, A. G. *Fette Seifen Anstrichm.* 58, 875, 1956
38. Bateman, G. M. and Sharp, P. F. *J. Agric. Res.* 36, 647, 1928
39. Cox, C. P. *J. Dairy Res.* 19, 72, 1952
40. Witnah, C. H., Rutz, W. D., and Freyer, H. C. *J. Dairy Sci.* 39, 1500, 1956
41. Søltoft, P. *On the consistency of mixtures of hardened fats.* Lewis. London, 1947
42. White, G. W. and Cakebread, S. H., *J. Fd. Technol.* 1, 73, 1966
43. Wood, F. W. *Society of Chemical Industry Monograph 27*, 1968
44. Charm, S. *Advances in Food Research.* 11, 356, 1962
45. Harper, J. C. *Fd. Technol.* 14, 557, 1960
46. Ward, A. G. and Westbrook, F. L. E. *J. Soc. Chem. Ind.* 67, 389, 1948
47. Ward, A. G. *Brit. J. Appl. Phys.* 1, 113, 1950
48. Price-Jones, J. *The Rheology of Honey*, in Scott-Blair, G. W. *Foodstuffs, their plasticity, fluidity and consistency.* North Holland, 1953
49. Eisenschitz, R. *Kolloidzeitschr.* 64, 184, 1933
50. Rutgers, R. *Getr. u. Mehl.* 8, 91, 1958
51. Whistler, R. L. and BeMiller, J. N. *Industrial Gums.* Academic Press, 1959
52. McDowell, R. H. *Alginates and their derivatives in water soluble gums and colloids.* Society of Chemical Industry Monograph. 24, 1966
53. Manuf. Gebrüder Haake, Berlin. Germany
54. Hoagland, S. (Ed.) *Rheology of surface coatings.* R. B. H. Dispersion Inc. N.J. 1946
55. Blackman, M. *Trans. Farad. Soc.* 44, 205, 1948
56. Eolkin, D. *Fd. Technol.* 11, 253, 1957
57. FIRA/NIRD Extruder. H. A. Gaydon & Co. Ltd. 93, Lansdowne Rd. Croydon, England. CR0 2BF
58. Claasens, J. W. *S. Afr. J. Agric. Sci.* 4, 457, 1958
59. Stefan, M. J. *Sitzber. Akad. Wiss. Wien. Math.-Naturw. Kl. Abt.* 11, 69, 713, 1874
60. Heiss, R. *J. Fd. Technol.* 13, 433, 1959
61. Autard, P. Society of Chemical Industry Monograph 27, *Rheology and Texture of Foodstuffs.* p. 236, 1968

62. Thomasos, F. I. and Wood, F. W. *J. Dairy. Res.* 31, 137, 1964
63. Muller, H. G. Society of Chemical Industry Monograph 27, *Rheology and Texture of Foodstuffs.* p. 183, 1968
64. Brown, P. Final Year Research Project. Procter Dept. Leeds University. 1969
65. Blackman, M. (*See* Matalou, R. *Foams* in *Flow properties of disperse systems* by Hermans, J. J. North Holland, 1953)
66. Elson, C. R. *B.F.M.I.R.A. Res. Rep.* No. 173, 1971
67. Steiner, E. H. *Confy. Prod.* 29, 439, 1963
68. Davies, J. G. *J. Dairy Res.* 8, 245, 1937
69. Haighton, A. J. *J. Amer. Oil Chem. Soc.* 36, 345, 1959
70. Eirich, F. R. *Rheology* Vol. 1. Academic Press, 1956
71. Rambaut, P. C., Bourland, C. T., Heidelbaugh, N. D., Huber, C. S., and Smith, M. C. *Fd. Technol.* 26, 1, 58–63, 1972
72. Muller, H. G. *Aspects of Dough Rheology* in Society of Chemical Industry Monograph 27, 1968
73. Funt, C. B., Lerchental, C. H., and Muller, H. G. *Rheological effects produced by additives to wheat flour dough* in Society of Chemical Industry Monograph 27, 1968
74. Stainsby, G. and Ward, A. G. *An absolute viscometer for measuring the viscoelastic properties of concentrated high polymer solutions.* In *Proc. Int. Cong. Rheol.* 1948. North Holland Pub. Co., 1949.
75. Hlynka, I. and Anderson. J. A. *Can. J. Technol.* 30, 198, 1952
76. Shelef, L. and Busso, D. *Rheol. Acta.* 3, 168, 1964
77. The Weissenberg Rheogoniometer. Sangamo Controls Ltd. Bognor Regis, Sussex, England
78. Criddle, D. W. *The Viscosity and elasticity of interfaces* in Eirich, F. R. *Rheology* Vol. 3. Academic Press, 1960
79. Tachibana, T. and Inokuchi, K. *J. Colloid Sci.* 8, 341, 1953
80. Michaud, F. *Ann. de Phys.* 19, 63, 1923
81. Saunders, P. R. and Ward, A. G. in *Rheology of Elastomers*, p. 45. Pergamon, 1958
82. Ward, A. G. and Cobbett, W. G. *Mechanical behaviour of dilute gelatin gels under large strains.* Society of Chemical Industry Monograph 27, 1968
83. Miller, M., Ferry, J. D., Schremp, F. W., and Eldridge, J. E. *J. Phys. Colloid Chem.* 55, 1387, 1951
84. Locken, L., Loska, S., and Schuey, W. *Farinograph Handbook*, American Association of Cereal Chemists, 1960
85. Muller, H. G. *Nature* (Lond.) 195, 4838, 235, 1962
86. Lerchental, C. H. and Muller, H. G. *C.S.T.* 12, 185, 1967
87. Shelef, L. and Busso, D. *Rheol. Acta.* 3. 168, 1964
88. Bale, R. and Muller, H. G. *J. Fd. Technol.* 5, 295, 1970
89. Shama, F. and Sherman, P. *An automated parallel-plate viscoelastometer for studying the rheological properties of solid food materials.* Society of Chemical Industry Monograph 27, 1968
90. Wade, P. *A texture meter for the measurement of biscuit hardness.* Society of Chemical Industry Monograph 27, 1968

91. Mohsenin, N. N. *Physical properties of plant and animal materials.* Vol. 1. Gordon and Breach, 1970
92. Bingham, E. C. and Wikoff, R. W. *J. Rheol.* 2, 395, 1931
93. White, G. W., Bell, A. V. and Berry, G. K. *J. Fd. Technol.* 2, 45, 1967
94. Zenz, F. A. and Othmer, D. F. *Fluidisation and fluid particle systems.* Reinhold, 1960
95. Matheson, G. L., Herbst, W. A., and Holt, P. H. *Ind. Eng. Chem.* 41, 1099, 1949

Index

Page numbers giving the definition of the entry are shown in *italics*.